AF652593

HISTORY OF HINDU MATHEMATICS

OTHER BOOKS PUBLISHED ON ANCIENT INDIAN SCIENCES

I. HINDU MATHEMATICS

1. **The Ganita-Sara-Sangraha of Mahaviracarya**, M Rangacarya
2. **The Hindu Art of Reckoning**, S R Benedict
3. **The Sulvasutras**, G Thibaut
4. **Karanaprakasa By Brahmadeva. With a Commentary and Theory of Numbers**, Ed. Sudhakara Dvivedi
5. **Bija Ganita. Mathematics of the Ancient Hindus**, E Starchey
6. **The Hindu-Arabic Numerals**, D. E. Smith & L. C. Karpinski
7. **Algebra, with Arithmetic and Mensuration from the Sanskrit of Brahmagupta and Bhaskara**, Translated by H. T. Colebrooke
8. **Early Hindu Geometry. The Science of the Sulba**, B. Datta
9. **Hindu Algebra**, H. T. Colebrooke
10. **History of Hindu Mathematics**, S. Datta, 2 Parts
11. **Ancient Indian Weights**, Edward Thomas

II. BIOLOGY, CHEMISTRY, GEOLOGY, Materia Medica

12. **The Vedic Gods as Figures of Biology**, V. G. Rele
13. **Materia Medica of the Hindus. Compiled from the Sanskrit Original Works**, G. King
14. **A History of Hindu Chemistry. From the Earliest Times to the Middle of the Sixteenth Century A.D.**, P. C. Ray
15. **Rasa-Jala-Nidhi. Ocean of Hindu Chemistry and Alchemy**, B. Mookerji, 4 Parts
16. **The Vedic Fathers of Geology**, N. B. Pavgee

III. HINDU MEDICINE AND SURGERY

17. **A History of Aryan Medical Sciences**, Bhagvat Sinh
18. **Antiquity of Hindoo Medicine. Along with Materia Medica and Therapeutics**, J. F. Royle
19. **An Interpretation of Ancient Hindu Medicine**, C. Chakraberty, 2 Parts
20. **Ancient Hindu Surgery. Surgical Instruments of the Hindus with a comparative study of the Greek, Roman, Arab, and the Modern European Surgeons**, G. Mukhopadhyaya, 2 Parts
21. **Studies in Medicine of Ancient India. Osteology or the Bones of the Human Body**, A. F. Rudolf Hoernle
22. **The Sushruta Samhita. An English Translation based on the Original Sanskrit Texts**, Kaviraj Kunja Lal, 5 Parts
23. **Classical Doctrine of Indian Medicine**, J Filliozat

IV. HINDU ASTRONOMY

24. **The Aryabhatia**, H Kern
25. **The Aryabhatti of Aryabhatta. An Ancient Indian Work on Mathematics and Astronomy**, Tr. W E Clark
26. **Siddhanta Siromani. A Treatise on Astronomy By Bhaskaracharya**, Ed. Pandit Muralidhara Jha
27. **Jyautisha Siddhanta Sangraha. A Collection of Ancient Hindu Astronomical Works.** P Dvivedi
28. **A Historical View of Hindu Astronomy. The Earliest Dawn of that Science in India to the Present**, J Bently
29. **Hindu Astronomy**, W Brennand
30. **Surya Siddhanta. A Text-book of Hindu Astronomy**, Tr. E. Burgess,
31. **The Pachasiddhantika. The Astronomical Work of Varaha Mihira**, G Thibaut & M Dvivedi
32. **Brihat Jataka of Varahamihira**, Tr. N. Chidambaram Aiyar

V. AEROSPACE TECHNOLOGY

33. **Vimana. Flying in Ancient India**, Ed. S Kapoor

VI. HISTORY OF HINDU SCIENCES

34. **Hindu Achievements in Exact Sciences.**, B. K. Sarkar
35. **Science and Society in India. Foundations of Planning**, K. Motwani, *Foreword by* Radhakamal Mukherjee

HISTORY OF HINDU MATHEMATICS

A SOURCE BOOKS

BY

BIBHUTIBHUSHAN DATTA

and

AVADHESH NARAYAN SINGH

IN TWO PARTS

PART II

Cosmo Publications

2011 INDIA

All rights reserved for this Edition.
No part of this publication may be reproduced, or stored in retrival system, or transmitted in any form or by any means without the prior written permission of Cosmo Publications.

HISTORY OF HINDU MATHEMATICS
set in 2 volumes

First Published by
Cosmo Publications, 2011

© Cosmo Publications, New Delhi
Moral and Intellectual Rights of the Authors are exerted

ISBN 978-81-307-1208-6 (set)
978-81-307-1613-8 (volume 2)

Published by
COSMO PUBLICATIONS
24-B, Ansari Road, Darya Ganj,
New Delhi-110 002, INDIA
email.: genesis@ndb.vsnl.net.in
cosmopublications@gmail.com

Printed and Bound in India

इदं नम ऋषिभ्यः पूर्वजेभ्यः पूर्वेभ्यः पथिकृद्भ्यः

(*RV*, x. 14. 25)

To the Seers, our Ancestors, the first Path-makers

PREFACE

The present work forms Part II of our History of Hindu Mathematics and is devoted to the history of Algebra in India. It is intended to be a source book, and the subject is treated topicwise. Under each topic are collected together and set forth in chronological order translations of relevant Sanskrit texts as found in the Hindu mathematical works. This plan necessitates a certain amount of repetition. But it shows to the reader at a glance the improvements made from century to century

To gather materials for the book we have examined all the published mathematical treatises of the Hindus as well as most of the important manuscripts available in Indian libraries, a list of the most important of which has already been included in Part I. We have great pleasure in once more expressing our thanks to the authorities of the libraries at Madras, Bangalore, Trivandrum, Tripunithura, Baroda, Jammu, and Benares, and those of the India Office (London) and the Asiatic Society of Bengal for supplying transcripts of manuscripts or sending them to us for consultation. We are indebted also to Dr. R. P. Paranjpye, Vice-Chancellor of the Lucknow University, for help in securing for our use several manuscripts or their transcripts from the State libraries in India and the India Office.

In translating Sanskrit texts we have tried to be as literal and faithful as possible without sacrificing the spirit of the original, in order to preserve which we have at a few places used literal translations of Sanskrit technical terms instead of modern terminology. For instance, we have used the term 'pulveriser' for the equation $ax+by=1$, and the term 'Square-nature' for the equation $Nx^2+c=y^2$.

The use of symbols—letters of the alphabet to denote unknowns—and equations are the foundations of the science of algebra. The Hindus were the first to make systematic use of the letters of the alphabet to denote unknowns. They were also the first to classify and make a detailed study of equations. Thus they may be said to have given birth to the modern science of algebra.

A portion of the subject matter of this book has been available to scholars through papers by various authors and through Colebrooke's *Algebra with Arithmetic and Mensuration from the Sanscrit of Brahmegupta and Bhascara*, but about half of it is being presented here for the first time. For want of space it has not been possible to give a detailed comparison of the Hindu achievements in Algebra with those of other nations. For this the reader is referred to the general works on the history of mathematics by Cantor, Smith, and Tropfke, to Dixon's *History of the Theory of Numbers* and to Neugebauer's *Mathematische Keilschrift-Texte*. A study of this book along with the above standard works will reveal to the reader the remarkable progress in algebra made by the Hindus at an early date. It will also show that we are indebted to the Hindus for the technique and the fundamental results of algebra just as we owe to them the place-value notation and the elements of our arithmetic.

We have pleasure in expressing our thanks to Mr. T. N. Singh and Mr. Ahmad Ali for help in correcting proofs and to Mr. R. D. Misra for preparing the index to this volume.

LUCKNOW
March, 1938

BIBHUTIBHUSAN DATTA
AVADHESH NARAYAN SINGH

CONTENTS

CHAPTER III

ALGEBRA

CONTENTS

CONTENTS

Chapter III

ALGEBRA

1. GENERAL FEATURES

Name for Algebra. The Hindu name for the science of algebra is *bîjagaṇita*. *Bîja* means "element" or "analysis" and *gaṇita* "the science of calculation." Thus *bîjagaṇita* literally means "the science of calculation with elements" or "the science of analytical calculation." The epithet dates at least as far back as the time of Pṛthûdakasvâmî (860) who used it. Brahmagupta (628) calls algebra *kuṭṭaka-gaṇita*, or simply *kuṭṭaka*.[1] The term *kuṭṭaka*, meaning "pulveriser", refers to a branch of the science of algebra dealing particularly with the subject of indeterminate equations of the first degree. It is interesting to find that this subject was considered so important by the Hindus that the whole science of algebra was named after it in the beginning of the seventh century. Algebra is also called *avyakta-gaṇita* or "the science of calculation with unknowns" (*avyakta*=unknown) in contradistinction to the name *vyakta-gaṇita* or "the science of calculation with knowns" (*vyakta*=known) for arithmetic including geometry and mensuration.

Algebra Defined. Bhâskara II (1150) has defined algebra thus:

"Analysis (*bîja*) is certainly the innate intellect assisted by the various symbols (*varṇa*), which, for the

[1]See Bibhutibhusan Datta, "The scope and development of the Hindu Gaṇita," *IHQ*, V, 1929, pp. 479-512; particularly pp. 489f.

instruction of duller intellects, has been expounded by the ancient sages who enlighten mathematicians as the sun irradiates the lotus; that has now taken the name algebra (*bījagaṇita*)."[1]

That algebraic analysis requires keen intelligence and sagacity has been observed by him on more than one occasion.

"Neither does analysis consist in symbols, nor are there different kinds of analyses; sagacity alone is analysis, for wide is imagination."[2]

"Analysis is certainly clear intelligence."[3]

"Or intelligence alone is analysis."[4]

In answer to the question, "if (unknown quantities) are to be discovered by intelligence alone what then is the need of analysis ?" he says:

"Because intelligence is certainly the real analysis; symbols are its helps. The innate intelligence which has been expressed for the duller intellects by the ancient sages, who enlighten mathematicians as the sun irradiates the lotus, with the help of various symbols, has now obtained the name of algebra."[5]

Thus, according to Bhâskara II, algebra may be defined as the science which treats of numbers expressed by means of symbols, and in which there is scope and primary need for intelligent artifices and ingenious devices.

Distinction from Arithmetic. What distinguishes algebra from arithmetic, according to the Hindus, will be found to some extent in their special names. Both deal with symbols. But in arithmetic the values of the symbols are *vyakta*, that is, known and definitely determinate,

[1] *BBi*, p. 99. [2] *BBi*, p. 49; *SiSi*, *Gola*, xiii. 5.
[3] L, p. 15; *SiŚi*, *Gola*, xiii. 3. [4] *BBi*, p. 49.
[5] *BBi*, p. 100.

while in algebra they are *avyakta*, that is, unknown, indefinite. The relation between these two branches of *gaṇita* is considered by Bhâskara II to be this:

"The science of calculation with unknowns is the source of the science of calculation with knowns."[1]

He has put it more explicitly and clearly thus:

"Algebra is similar to arithmetic in respect of rules (of fundamental operations) but appears as if it were indeterminate. It is not indeterminate to the intelligent; it is certainly not sixfold,[2] but manifold."[3]

The true distinction between arithmetic and algebra, besides that of symbols employed, lies, in the opinion of Bhâskara II, in the demonstration of the rules. He remarks:

"Mathematicians have declared algebra to be computation attended with demonstration: else there would be no distinction between arithmetic and algebra."[4]

The truth of this dictum is evident in the treatment of the *guṇa-karma* in the *Lîlâvatî* and the *madhyamâharaṇa* in the *Bîjagaṇita*. Both are practically treatments of problems involving the quadratic equation. But whereas in the former are found simply the applications of the well-known formulæ for the solution of such equations, in the latter is described also the *rationale* of those formulæ. Similarly we sometimes find included in treatises on arithmetic problems whose solutions require formulæ demonstrated in books on algebra. The method of demonstration has been stated to be "always of two kinds: one geometrical (*kṣetragata*) and

[1] *BBi*, p. 1.

[2] The reference is to the six fundamental operations recognised in algebra as well as to the six subjects of treatment which are essential to analysis.

[3] *L*, p. 15.

[4] *BBi*, p. 127.

the other symbolical (*râśigata*)."[1] We do not know who was the first in India to use geometrical methods for demonstrating algebraical rules. Bhâskara II (1150) ascribes it to "ancient teachers."[2]

Importance of Algebra. The early Hindus regarded algebra as a science of great importance and utility. In the opening verses of his treatise[3] on algebra Brahmagupta (628) observes:

"Since questions can scarcely be known (*i.e.*, solved) without algebra, therefore, I shall speak of algebra with examples.

"By knowing the pulveriser, zero, negative and positive quantities, unknowns, elimination of the middle term, equations with one unknown, factum and the Square-nature, one becomes the learned professor (*âcârya*) amongst the learned."[4]

Similarly Bhâskara II writes:

"What the learned calculators (*sâṁkhyâḥ*) describe as the originator of intelligence, being directed by a wise being (*satpuruṣa*) and which alone is the primal cause (*bîja*) of all knowns (*vyakta*), I venerate that Invisible God as well as that Science of Calculation with Unknowns... Since questions can scarcely be solved without the reasoning of algebra—not at all by those of dull perceptions—I shall speak, therefore, of the operations of analysis."[5]

[1] *BBi*, p. 125. [2] *BBi*, p. 127.

[3] Forming chapter xviii of his *Brâhma-sphuṭa-siddhânta.*

[4] *BrSpSi*, xviii. 1-2.

[5] In the first part of this passage every principal term has been used with a double significance. The term *sâṁkhyâḥ* (literally, "expert calculators") signifies the "Sâṁkhya philosophers" in one sense, "mathematicians" in the other; *satpuruṣa* "the self-existent being of the Sâṁkhya philosophy" or "a wise mathematician"; *vyakta* "manifested universe" or "the science of calculation with knowns."

Nârâyaṇa (1350) remarks :

"I adore that Brahma, also that science of calculation with the unknown, which is the one invisible root-cause of the visible and multiple-qualitied universe, also of multitudes of rules of the science of calculation with the known."[1]

"As out of Him is derived this entire universe, visible and endless, so out of algebra follows the whole of arithmetic with its endless varieties (of rules). Therefore, I always make obeisance to Siva and also to *(avyakta-) gaṇita* (algebra)."[2]

He adds :

"People ask questions whose solutions are not to be found by arithmetic; but their solutions can generally be found by algebra. Since less intelligent men do not succeed in solving questions by the rules of arithmetic, I shall speak of the lucid and easily intelligible rules of algebra."[3]

Scope of Algebra. The science of algebra is broadly divided by the Hindus into two principal parts. Of these the most important one deals with analysis (*bîja*). The other part treats of the subjects which are essential for analysis. They are: the laws of signs, the arithmetic of zero (and infinity), operations with unknowns, surds, the pulveriser (or the indeterminate equation of the first degree), and the Square-nature (or the so-called Pellian equation). To these some writers add concurrence and dissimilar operations, while others include them in arithmetic.[4] At the end of the first section of his treatise on algebra Bhâskara II is found to have

[1] *NBi*, I, R. 1. [2] *NBi*, II, R. 1.

[3] *NBi*, I, R. 5-6.

[4] All writers, except Brahmagupta and Śrîpati, are of the latter opinion.

observed as follows :

"(The section of) this science of calculation which is essential for analysis has been briefly set forth. Next I shall propound analysis, which is the source of pleasure to the mathematician."[1]

Analysis is stated by all to be of four kinds, for equations are classified into four varieties (*vide infra*). Thus each class of equations has its own method of analysis.

Origin of Hindu Algebra. The origin of Hindu algebra can be definitely traced back to the period of the *Śulba* (800-500 B.C.) and the *Brâhmaṇa* (*c.* 2000 B.C.). But it was then mostly geometrical.[2] The geometrical method of the transformation of a square into a rectangle having a given side, which is described in the important *Śulba* is obviously equivalent to the solution of a linear equation in one unknown, *viz.*,

$$ax = c^2.$$

The quadratic equation has its counterpart in the construction of a figure (an altar) similar to a given one but differing in area from it by a specified amount. The usual method of solving that problem was to increase the unit of measure of the linear dimensions of the figure. One of the most important altars of the obligatory Vedic sacrifices was called the *Mahâvedi* (the Great Altar). It has been described to be of the form of an isosceles trapezium whose face is 24 units long, base 30 and altitude 36. If x be the enlarged unit of measure taken in increasing the size of the altar by m units of area, we must have

$$36x \times \frac{(24x + 30x)}{2} = 36 \times \frac{(24 + 30)}{2} + m,$$

[1] *BBi*, p. 43.

[2] Bibhutibhusan Datta, *The Science of the Sulba*, Calcutta, 1932.

or $$972x^2 = 972 + m.$$

Therefore $$x = \sqrt{1 + \frac{m}{972}}.$$

If m be put equal to 972 $(n-1)$, so that the area of the enlarged altar is n times its original area, we get

$$x = \sqrt{n},$$

some particular cases of which are described in the *Śulba*. The particular cases, when $n = 14$ or $14\frac{3}{7}$, are found as early as the *Śatapatha Brâhmaṇa*[1] (*c.* 2000 B.C.).

The most ancient and primitive form of the "Fire-altar for the sacrifices to achieve special objects" was the *Śyenacit* (or "the altar of the form of the falcon").

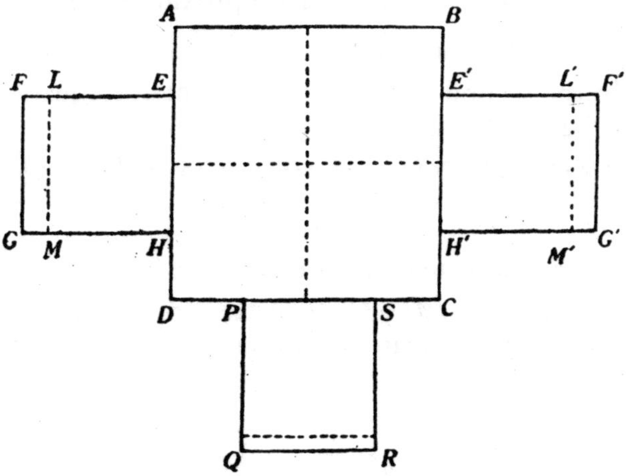

Fig. 1.

Its body ($ABCD$) consists of four squares of one square *puruṣa* each; each of its wings ($EFGH$, $E'F'G'H'$) is a rectangle of one *puruṣa* by one *puruṣa* and a *prâdeśa* (=1/10 of a *puruṣa*). This Fire-altar was enlarged in

[1] *ŚBr*, X. 2. 3. 7ff.

two ways: *first*, in which all the constituent parts were affected in the same proportion; *second*, in which the breadth of the portions *LFGM* and *L'F'G'M'* of the wings were left unaffected. If x be the enlarged unit for enlargement in the first case we shall have to solve the quadratic equation

$$2x \times 2x + 2\left\{x \times \left(x + \frac{x}{5}\right)\right\} + x \times \left(x + \frac{x}{10}\right) = 7\tfrac{1}{2} + m,$$

where m denotes the increment of the Fire-altar in size.

Therefore $$x^2 = 1 + \frac{2m}{15}.$$

In particular, when $m = 94$, we shall have

$$x^2 = 13\tfrac{8}{15} = 14 \text{ (approximately)},$$

which occurs in the *Śatapatha Brâhmaṇa*.

In the second case of enlargement the equation for x will be

$$2x \times 2x + 2\{x \times (x + \tfrac{1}{5})\} + x \times (x + \tfrac{1}{10}) = 7\tfrac{1}{2} + m,$$

or $$7x^2 + \tfrac{1}{2}x = 7\tfrac{1}{2} + m,$$

which is a complete quadratic equation.

The problem of altar construction gave rise also to certain indeterminate equations of the second degree such as,

(1) $$x^2 + y^2 = z^2,$$

(2) $$x^2 + a^2 = z^2;$$

and simultaneous indeterminate equations of the type

$$ax + by + cz + dw = p,$$
$$x + y + z + w = q.$$

Further particulars about these equations will be given later on.

2. TECHNICAL TERMS

Coefficient. In Hindu algebra there is no systematic use of any special term for the coefficient. Ordinarily, the power of the unknown is mentioned when the reference is to the coefficient of that power. In explanation of similar use by Brahmagupta his commentator Pṛthûdakasvâmî writes "the number (*aṅka*) which is the coefficient of the square of the unknown is called the 'square' and the number which forms the coefficient of the (simple) unknown is called 'the unknown quantity.'"[1] However, occasional use of a technical term is also met with. Brahmagupta once calls the coefficient *saṁkhyâ*[2] (number) and on several other occasions *guṇaka*, or *guṇakâra* (multiplier).[3] Pṛthûdakasvâmî (860) calls it *aṅka*[4] (number) or *prakṛti* (multiplier). These terms reappear in the works of Śrîpati (1039)[5] and Bhâskara II (1150).[6] The former also used *rûpa* for the same purpose.[7]

Unknown Quantity. The unknown quantity was called in the *Sthânâṅga-sûtra*[8] (before 300 B.C.) *yâvat-tâvat* (as many as or so much as, meaning an arbitrary quantity). In the so-called Bakhshâlî treatise, it was called *yadṛcchâ*, *vâñchâ* or *kâmika* (any desired quantity).[9] This term was originally connected with the Rule of False Position.[10] Âryabhaṭa I (499)

[1] *BrSpSi*, xviii. 44 (*Com*).
[2] *BrSpSi*, xviii. 63.
[3] *BrSpSi*, xviii. 64, 69-71.
[4] *BrSpSi*, xviii. 44 (*Com*).
[5] *SiŚe*, xiv. 33-5.
[6] *BBi*, pp. 33-4.
[7] *SiŚe*, xiv. 19.
[8] *Sûtra* 747; *cf.* Bibhutibhusan Datta, "The Jaina School of Mathematics," *BCMS*, XXI, 1929, pp. 115-145; particularly pp. 122-3.
[9] *BMs*, Folios 22, verso; 23, recto & verso.
[10] Bibhutibhusan Datta, "The Bakshshâlî Mathematics," *BCMS*, XXI, pp. 1-60; particularly pp. 26-8, 66.

calls the unknown quantity *gulikâ* (shot). This term strongly leads one to suspect that the shot was probably then used to represent the unknown. From the beginning of the seventh century the Hindu algebraists are found to have more commonly employed the term *avyakta* (unknown).[1]

Power. The oldest Hindu terms for the power of a quantity, known or unknown, are found in the *Uttarâdhyayana-sûtra* (*c.* 300 B.C. or earlier).[2] In it the second power is called *varga* (square), the third power *ghana* (cube), the fourth power *varga-varga* (square-square), the sixth power *ghana-varga* (cube-square), and the twelfth power *ghana-varga-varga* (cube-square-square), using the multiplicative instead of the additive principle. In this work we do not find any method for indicating odd powers higher than the third. In later times, the fifth power is called *varga-ghana-ghâta* (product of cube and square, *ghâta*=product), the seventh power *varga-varga-ghana-ghâta* (product of square-square and cube) and so on. Brahmagupta's system of expressing powers higher than the fourth is scientifically better. He calls the fifth power *pañca-gata* (literally, raised to the fifth), the sixth power *ṣaḍ-gata* (raised to the sixth); similarly the term for any power is coined by adding the suffix *gata* to the name of the number indicating that power.[3] Bhâskara II has sometimes followed it consistently for the powers one and upwards.[4] In the *Anuyogadvâra-sûtra*[5], a work written before the commencement of the Christian Era, we find certain interesting terms for higher powers, integral as well as fractional, particularly successive squares (*varga*) and square-roots (*varga-mûla*). According to it the term *prathama-varga*

[1] *BrSpSi*, xviii. 2, 41; *SiŚe*, xiv. 1-2; *BBi*, pp. 7ff.
[2] Chapter xxx, 10, 11.
[3] *BrSpSi*, xviii. 41, 42.
[4] *BBi*, p. 56.
[5] *Sûtra* 142.

(first square) of a quantity, say a, means a^2; *dvitîyavarga* (second square) $= (a^2)^2 = a^4$; *tṛtîya-varga* (third square) $= ((a^2)^2)^2 = a^8$; and so on. In general,

$$n\text{th } varga \text{ of } a = a^{2\times 2\times 2\times \ldots \text{ to } n \text{ terms}} = a^{2^n}.$$

Similarly, *prathama-varga-mûla* (first square-root) means $\sqrt{a}$; *dvitîya-varga-mûla* (second square-root) $=\sqrt{(\sqrt{a})}$ $= a^{1/4}$; and, in general,

$$n\text{th } varga\text{-}m\hat{u}la \text{ of } a = a^{1/2^n}.$$

Again we find the term *tṛtîya-varga-mûla-ghana* (cube of the third square-root) for $(a^{1/2^3})^3 = a^{3/8}$.

The term *varga* for "square" has an interesting origin in a purely concrete concept. The Sanskrit word *varga* literally means "rows," or "troops" (of similar things). Its application as a mathematical term originated in the graphical representation of a square, which was divided into as many *varga* or troops of small squares, as the side contained units of some measure.[1]

Equation. The equation is called by Brahmagupta (628) *sama-karaṇa*[2] or *samî-karaṇa*[3] (making equal) or more simply *sama*[4] (equation). Pṛthûdakasvâmî (860) employs also the term *sâmya*[5] (equality or equation); and Śrîpati (1039) *sadṛśî-karaṇa*[6] (making similar). Nârâyaṇa (1350) uses the terms *samî-karaṇa*, *sâmya* and *samatva* (equality).[7] An equation has always two *pakṣa* (side). This term occurs in the works of

[1] G. Thibaut, *Sulba-sûtras*, p. 48. Compare also Bibhutibhusan Datta, "On the origin of the Hindu terms for root," *Amer. Math. Mon.*, XXXVIII, 1931, pp. 371-6.

[2] *BrSpSi*, xviii. 63.

[3] *BrSpSi*, xviii, subheading for the section on equations.

[4] *BrSpSi*, xviii. 43.

[5] *BrSpSi*, xii. 66 (*Com*).

[6] *SiSe*, xiv. 1.

[7] *NBi*, II, R. 2-3.

Srîdhara (*c.* 750), Padmanâbha[1] and others.[2]

Absolute Term. In the Bakhshâlî treatise[3] the absolute term is called *dṛśya* (visible). In later Hindu algebras it has been replaced by a closely allied term *rûpa*[4] (appearance), though it continued to be employed in treatises on arithmetic.[5] Thus the true significance of the Hindu name for the absolute term in an algebraic equation is obvious. It represents the visible or known portion of the equation while its other part is practically invisible or unknown.

3. SYMBOLS

Symbols of Operation. There are no special symbols for the fundamental operations in the Bakhshâlî work. Any particular operation intended is ordinarily indicated by placing the tachygraphic abbreviation, the initial syllable of a Sanskrit word of that import, after, occasionally before, the quantity affected. Thus the operation of addition is indicated by *yu* (an abbreviation from *yuta*, meaning added), subtraction by + which is very probably from *kṣa* (abbreviated from *kṣaya*, diminished), multiplication by *gu* (from *guṇa* or *guṇita*, multiplied) and division by *bhâ* (from *bhâga* or *bhâjita*, divided). Of these again, the most systematically employed abbreviation is that for the operation of subtraction and next comes that of division. In the case of the other two operations the indicatory words

[1] The algebras of Śrîdhara and Padmanâbha are not available now. But the term occurs in quotations from them by Bhâskara II (*BBi*, pp. 61, 67).

[2] *BrSpSi*, xviii. 43 (*Com*); *SiŚe*, xiv. 14, 20; *BBi*, pp. 43-4.

[3] *BMs*, Folio 23, verso; Folio 70, recto and verso (c).

[4] *BrSpSi*, xviii. 43-4; *SiŚe*. xiv, 14, 19; etc.

[5] *Triś*, pp. 11, 12.

are often written in full, or altogether omitted. In the latter case, the particular operations intended to be carried out are understood from the context. We take the following instances from the Bakhshâlî Manuscript:

(i)[1] $\begin{array}{cc}0 & 5\\ 1 & 1\end{array}$ *yu* means $\frac{x}{1}+\frac{5}{1}$, and $\begin{array}{c}11\\ 1\end{array}$ *yu* $\begin{array}{c}5\\ 1\end{array}$ means $\frac{11}{1}+\frac{5}{1}$.

(ii)[2] $\left\| \begin{array}{cccccccc}3 & 3 & 3 & 3 & 3 & 3 & 3 & 10\, gu\\ 1 & 1 & 1 & 1 & 1 & 1 & 1 & 1\end{array} \right|$ means $3\times3\times3\times3\times3\times3\times3\times10$.

(iii)[3] $\begin{array}{c}0\\ 1\end{array} \left\| \begin{array}{cc}1 & 3\\ 1 & 2\end{array} \right\| \begin{array}{cc}2 & 5\\ 1 & 2+\end{array} \left\| \begin{array}{ccc}3 & gu & 7\\ 1 & & 2+\end{array} \right\| \begin{array}{ccc}4 & gu & 9\\ 1 & & 2+\end{array} \right\|$

means

$$x(1+\tfrac{3}{2})+\left\{2x(1+\tfrac{3}{2})-\frac{5x}{2}\right\}+\left\{3x(1+\tfrac{3}{2})-\frac{7x}{2}\right\}+\left\{4x(1+\tfrac{3}{2})-\frac{9x}{2}\right\}.$$

(iv)[4] $\left| \begin{array}{cccc}1 & 1 & 1 & 1\, bh\hat{a}\\ 1+ & 1 & 1 & 1\\ 2 & 3 & 4+ & 6\end{array} \right\| \begin{array}{c}36\\ \\ 1\end{array} \right|$ means $\dfrac{36}{(1-\frac{1}{2})(1+\frac{1}{3})(1-\frac{1}{4})(1+\frac{1}{6})}$.

(v)[5] $\left| \begin{array}{c}40\ bh\hat{a}\\ 1\end{array} \right\| \begin{array}{c}160\\ 1\end{array} \left| \begin{array}{c}13\\ 1\\ 2\end{array} \right\|$ means $\dfrac{160}{40}\times13\tfrac{1}{2}$.

In later Hindu mathematics the symbol for subtraction is a dot, occasionally a small circle, which is placed above the quantity, so that $\dot{7}$ or $\mathring{7}$ means -7; other operations are represented by simple juxtaposition.

[1] Folio 59, recto.
[2] Folio 47, recto.
[3] Folio 25, verso. The beginning and end of this illustration are mutilated but the restoration is certain.
[4] Folio 13, verso.
[5] Folio 42, recto.

Bhâskara II (1150) says, "Those (known and unknown numbers) which are negative should be written with a dot (*bindu*) over them."[1] A similar remark occurs in the algebra of Nârâyaṇa (1350).[2] Their silence about symbols of other fundamental operations proves their non-existence.

Origin of Minus Sign. The origin of · or ° as the minus sign seems to be connected with the Hindu symbol for the zero, o. It is found to have been used as the sign of emptiness or omission in the early Bakhshâlî treatise as well as in the later treatises on arithmetic (*vide infra*).[3] It is placed over the number affected in order to distinguish it from its use in a purely numerical significance when it is placed beside the number. The origin of the Bakhshâlî minus sign (+) has been the subject of much conjecture. Thibaut suggested its possible connection with the supposed Diophantine negative sign ⋔ (reversed ψ, tachygraphic abbreviation for λεiψis meaning wanting). Kaye believes it. The Greek sign for minus, however, is not ⋔, but ↑. It is even doubtful if Diophantus did actually use it; or whether it is as old as the Bakhshâlî cross.[4] Hoernle[5] presumed the Bakhshâlî minus sign to be the abbreviation *ka* of the Sanskrit word *kanita*, or *nu* (or *nû*) of *nyûna*, both of which mean diminished and both of which abbreviations in the Brâhmî characters would be denoted by a cross. Hoernle was right, thinks Datta,[6] so far as he sought for the origin of + in a tachygraphic abbreviation of some Sanskrit word. But as neither the word *kanita* nor *nyûna* is found to have been used in the Bakhshâlî work in connection with the subtractive

[1] *BBi*, p. 2.
[2] *NBi*, I, R. 7.
[3] p. 16.
[4] *Cf.* Smith, *History*, II, p. 396.
[5] *IA*, XVII, p. 34.
[6] Datta, *Bakh. Math.*, (*BCMS*, XXI), pp. 17-8.

operation, Datta finally rejects the theory of Hoernle and believes it to be the abbreviation *kṣa*, from *kṣaya* (decrease) which occurs several times, indeed, more than any other word indicative of subtraction. The sign for *kṣa*, whether in the Brâhmî characters or in Bakhshâlî characters, differs from the simple cross (+) only in having a little flourish at the lower end of the vertical line. The flourish seems to have been dropped subsequently for convenient simplification.

Symbols for Powers and Roots. The symbols for powers and roots are abbreviations of Sanskrit words of those imports and are placed after the number affected. Thus the square is represented by *va* (from *varga*), cube by *gha* (from *ghana*), the fourth power by *va-va* (from *varga-varga*), the fifth power by *va-gha-ghâ* (from *varga-ghana-ghâta*), the sixth power by *gha-va* (from *ghana-varga*), the seventh power by *va-va-gha-ghâ* (from *varga-varga-ghana-ghâta*) and so on. The product of two or more unknown quantities is indicated by writing *bhâ* (from *bhâvita*, product) after the unknowns with or without interposed dots ; e.g., *yâva-kâgha-bhâ* or *yâvakâghabhâ* means $(y\hat{a})^2\,(k\hat{a})^3$. In the Bakhshâlî treatise the square-root of a quantity is indicated by writing after it *mû*, which is an abbreviation for *mûla* (root). For instance,[1]

11	*yu*	5	*mû*	4
1		1		1

means $\sqrt{11 + 5} = 4$

and

11	7+		*mû*	2
1	1			1

means $\sqrt{11 - 7} = 2$.

In other treatises the symbol of the square-root is *ka* (from *karaṇî*, root or surd) which is usually placed before the quantity affected. For example,[2] *ka* 9 *ka* 450

[1] Folio 59, recto; compare also folio 67, verso.

[2] *BBi*, p. 15.

ka 75 *ka* 54 means $\sqrt{9}+\sqrt{450}+\sqrt{75}+\sqrt{54}$.

Symbols for Unknowns. In the Bakhshâlî treatise there is no specific symbol for the unknown. Consequently its place in an equation is left vacant and to indicate it vividly the sign of emptiness is put there. For instance,[1]

0	2	3	4	*dṛśya* 200
1	1	1	1	1

means $x+2x+3x+4x=200$.

The use of the zero sign to mark a vacant place is found in the arithmetical treatises of later times when the Hindus had a well-developed system of symbols for the unknowns. Thus we find in the *Triśatikā*[2] of Srîdhara (*c.* 750) the following statement of an arithmetical progression whose first term (*âdiḥ*) is 20, number of terms (*gacchaḥ*) 7, sum (*gaṇitaṁ*) 245 and whose common difference (*uttaraḥ*) is unknown:

| *âdiḥ* 20 | *u* 0 | *gacchaḥ* 7 | *gaṇitaṁ* 245 |

This use of the zero sign in arithmetic was considered necessary as algebraic symbols could not be used there. Lack of an efficient symbolism is bound to give rise to a certain amount of ambiguity in the representation of an algebraic equation especially when it contains more than one known. For instance, in[3]

0 5 *yu* *mû* 0 1 1 1	*sa*	0 7+ *mû* 0 1 1 1

which means

$$\sqrt{x+5}=s \text{ and } \sqrt{x-7}=t,$$

different unknowns have to be assumed at different vacant places.

[1] *BMs*, Folio 22, verso. [2] *Triś*, p. 29.
[3] *BMs*, Folio 59, recto.

To avoid such ambiguity, in one instance which contains as many as five unknowns, the abbreviations of ordinal numbers, such as *pra* (from *prathama*, first), *dvi* (from *dvitîya*, second), *tṛ* (from *tṛtîya*, third), *ca* (from *caturtha*, fourth) and *paṁ* (from *pañcama*, fifth), have been used to represent the unknowns; *e.g.*,[1]

9 *pra*	7 *dvi*	10 *tṛ*	8 *ca*	11 *paṁ*	yutam jâtaṁ pratyaika-(kramena)
7 *dvi*	10 *tṛ*	8 *ca*	11 *paṁ*	9 *pra*	16|17|18|19|20

which means

$x_1(=9)+x_2(=7)=16;\ x_2(=7)+x_3(=10)=17;$
$x_3(=10)+x_4(=8)=18;\ x_4(=8)+x_5(=11)=19;$
$x_5(=11)+x_1(=9)=20.$

Âryabhaṭa I (499) very probably used coloured shots to represent unknowns. Brahmagupta (628) mentions *varṇa* as the symbols of unknowns.[2] As he has not attempted in any way to explain this method of symbolism, it appears that the method was already very familiar. Now, the Sanskrit word *varṇa* means "colour" as well as "letters of the alphabet," so that, in later times, the unknowns are generally represented by letters of the alphabet or by means of various colours such as *kâlaka* (black), *nîlaka* (blue), etc. Again in the latter case, for simplification, only initial letters of the names are generally written. Thus Bhâskara II (1150) observes, "Here (in algebra) the initial letters of (the names of) knowns and unknowns should be written for implying them."[3] It has been stated before that at one time the unknown quantity was called *yâvat-tâvat* (as many

[1] Folio 27, verso.
[2] *BrSpSi*, xviii. 2, 42, 51, etc.
[3] *BBi*, p. 2; see also *NBi*, I, R. 7.

as, so much as). In later times this name, or its abbreviation *yâ*, is used for the unknown. According to the celebrated Sanskrit lexicographer Amarasiṁha (*c.* 400 A.D.), *yâvat-tâvat* denotes measure or quantity (*mâna*). He had probably in view the use of that term in Hindu algebra to denote "the measure of an unknown" (*avyakta mâna*). In the case of more unknowns, it is usual to denote the first by *yâvat-tâvat* and the remaining ones by alphabets or colours. Pṛthûdakasvâmî (860) says:

"In an example in which there are two or more unknown quantities, colours such as *yâvat-tâvat*, etc., should be assumed for their values."[1]

He has, indeed, used the colours *kâlaka* (black), *nîlaka* (blue), *pîtaka* (yellow) and *harîtaka* (green).

Śrîpati (1039) writes:

"*Yâvat-tâvat* (so much as) and colours such as *kâlaka* (black), *nîlaka* (blue), etc., should be assumed for the unknowns."[2]

Bhâskara II (1150) says:

"*Yâvat-tâvat* (so much as), *kâlaka* (black), *nîlaka* (blue), *pîta* (yellow), *lohita* (red) and other colours have been taken by the venerable professors as notations for the measures of the unknowns, for the purpose of calculating with them."[3]

"In those examples where occur two, three or more unknown quantities, colours such as *yâvat-tâvat*, etc., should be assumed for them. As assumed by the previous teachers, they are: *yâvat-tâvat* (so much as), *kâlaka* (black), *nîlaka* (blue), *pîtaka* (yellow), *lohitaka* (red), *harîtaka* (green), *śvetaka* (white), *citraka*

[1] *BrSpSi*, xviii. 51 (*Com*). [2] *SiŚe*, xiv. 2.
[3] *BBi*, p. 7.

(variegated), *kapilaka* (tawny), *piṅgalaka* (reddish-brown), *dhûmraka* (smoke-coloured), *pâtalaka* (pink), *śavalaka* (spotted), *śyâmalaka* (blackish), *mecaka* (dark blue), etc. Or the letters of alphabets beginning with *ka*, should be taken as the measures of the unknowns in order to prevent confusion."[1]

The same list with a few additional names of colours appears in the algebra of Nârâyaṇa.[2] This writer has further added,

"Or the letters of alphabets (*varṇa*) such as *ka*, etc., or the series of flavours such as *madhura* (sweet), etc., or the names of dissimilar things with unlike initial letters, are assumed (to represent the unknowns)."

Bhâskara II occasionally employs also the tachygraphic abbreviation of the names of the unknown quantities themselves in order to represent them in an equation. For example,[3] in the following

5 *mâ*	1 *nî*	1 *mu*	1 *va*
1 *mâ*	7 *nî*	1 *mu*	1 *va*
1 *mâ*	1 *nî*	97 *mu*	1 *va*
1 *mâ*	1 *nî*	1 *mu*	2 *va*

mâ stands for *mâṇikya* (ruby), *nî* for (*indra-*)*nîla* (sapphire), *mu* for *muktâphala* (pearl) and *va* for (*sad*)*vajra* (diamond). He has observed in this connection thus:

"(The maxim), 'colours such as *yâvat-tâvat*, etc., should be assumed for the unknowns,' gives (only) one method of implying (them). Here, denoting them

[1] *BBi*, pp. 76f.

[2] *NBi*, I, R. 17-8. These verses have been quoted by Muralidhara Jha in his edition of the *Bîjagaṇita* of Bhâskara II (p. 7, footnote 5).

[3] *BBi*, p. 50; compare also p. 28.

by names, the equations may be formed by the intelligent (calculator)."

It should be noted that *yâvat-tâvat* is not a *varṇa* (colour or letter of alphabet). So in its inclusion in the lists of *varṇa*, as found enumerated in the Hindu algebras—though apparently anomalous—we find the persistence of an ancient symbol which was in vogue long before the introduction of colours to represent unknowns. To avoid the anomaly Muralidhara Jha[1] has suggested the emendation *yâvakastâvat* (*yâvaka* and also; *yâvaka* = red) in the place of *yâvat-tâvat*, as found in the available manuscripts. He thinks that being misled by the old practice, the expression *yâvakastâvat* was confused by copyists with *yâvat-tâvat*. In support of this theory it may be pointed out that *yâvaka* is found to have been sometimes used by Pṛthû lakasvâmî to represent the unknown.[2] Bhâskara II has once used simply *yâvat*.[3] Nârâyana used it on several occasions. The origin of the use of names of colours to represent unknowns in algebra is very probably connected with the ancient use of differently coloured shots for the purpose.

4. LAWS OF SIGNS

Addition. Brahmagupta (628) says:

"The sum of two positive numbers is positive, of two negative numbers is negative; of a positive and a negative number is their difference."[4]

Mahâvîra (850):

"In the addition of a positive and a negative number

[1] See the Preface to his edition of Bhâskara's *Bîjagaṇita*.
[2] *BrSpSi*, xii. 15 (*Com*); xii. 18 (*Com*).
[3] *BBi*, p. 50.
[4] *BrSpSi*, xviii. 30.

(the result) is (their) difference. The addition of two positive or two negative numbers (gives) as much positive or negative numbers respectively."[1]

Srîpati (1039):

"In the addition of two negative or two positive numbers the result is their sum; the addition of a positive and a negative number is their difference."[2]

"The sum of two positive (numbers) is positive; of two negative (numbers) is negative; of a positive and a negative is their difference and the sign of the difference is that of the greater; of two equal positive and negative (numbers) is zero."[3]

Bhâskara II (1150):

"In the addition of two negative or two positive numbers the result is their sum; the sum of a positive and a negative number is their difference."[4]

Nârâyaṇa (1350):

"In the addition of two positive or two negative numbers the result is their sum; but of a positive and a negative number, the result is their difference; subtracting the smaller number from the greater, the remainder becomes of the same kind as the latter."[5]

Subtraction. Brahmagupta writes:

"From the greater should be subtracted the smaller; (the final result is) positive, if positive from positive, and negative, if negative from negative. If, however, the greater is subtracted from the less, that difference is reversed (in sign), negative becomes positive and positive becomes negative. When positive is to be subtracted from negative or negative from positive,

[1] *GSS*, i. 50-1.
[2] *SiSe*, xiv. 3.
[3] *SiŚe*, iii. 28.
[4] *BBi*, p. 2.
[5] *NBi*, I, R. 8.

then they must be added together."[1]

Mahâvîra:

"A positive number to be subtracted from another number becomes negative and a negative number to be subtracted becomes positive."[2]

Śrîpati:

"A positive (number) to be subtracted becomes negative, a negative becomes positive; (the subsequent operation is) addition as explained before."[3]

Bhâskara II:

"A positive (number) while being subtracted becomes negative and a negative becomes positive; then addition as explained before."[4]

Nârâyaṇa:

"Of the subtrahend affirmation becomes negation and negation affirmation; then addition as described before."[5]

Multiplication. Brahmagupta says:

"The product of a positive and a negative (number) is negative; of two negatives is positive; positive multiplied by positive is positive."[6]

Mahâvîra:

"In the multiplication of two negative or two positive numbers the result is positive; but it is negative in the case of (the multiplication of) a positive and a negative number."[7]

Śrîpati:

"On multiplying two negative or two positive

[1] *BrSpSi*, xviii. 31-2.
[2] *GSS*, i. 51.
[3] *SiŚe*, xiv. 3.
[4] *BBi*, p. 3.
[5] *NBi*, I, R. 9.
[6] *BrSpSi*, xviii. 33.
[7] *GSS*, i. 50.

numbers (the product is) positive; in the multiplication of positive and negative (the result is) negative."[1]

Bhâskara II:

"The product of two positive or two negative (numbers) is positive; the product of positive and negative is negative."[2]

The same rule is stated by Nârâyaṇa.[3]

Division. Brahmagupta states:

"Positive divided by positive or negative divided by negative becomes positive. But positive divided by negative is negative and negative divided by positive remains negative."[4]

Mahâvîra:

"In the division of two negative or two positive numbers the quotient is positive, but it is negative in the case of (the division of) positive and negative."[5]

Śrîpati:

"On dividing negative by negative or positive by positive, (the quotient) will be positive, (but it will be) otherwise in the case of positive and negative."[6]

Bhâskara II simply observes: "In the case of division also, such are the rules (*i.e.*, as in the case of multiplication)."[7] Similarly Nârâyaṇa remarks, "What have been implied in the case of multiplication of positive and negative numbers will hold also in the case of division."[8]

Evolution and Involution. Brahmagupta says:

"The square of a positive or a negative number is

[1] *SiŚe*, xiv. 4.
[2] *BBi*, p. 3.
[3] *NBi*, I, R. 9.
[4] *BrSpSi*, xviii. 34.
[5] *GSS*, i. 50.
[6] *SiŚe*, xiv. 4.
[7] *BBi*, p. 3.
[8] *NBi* I, R. 10.

positive. . . . The (sign of the) root is the same as was that from which the square was derived."[1]

As regards the latter portion of this rule the commentator Pṛthûdakasvâmî (860) remarks, "The square-root should be taken either negative or positive, as will be most suitable for subsequent operations to be carried on."

Mahâvîra:

"The square of a positive or of a negative number is positive: their square-roots are positive and negative respectively. Since a negative number by its own nature is not a square, it has no square-root."[2]

Srîpati:

"The square of a positive and a negative number is positive. It will become what it was in the case of the square-root. A negative number by itself is non-square, so its square-root is unreal; so the rule (for the square-root) should be applied in the case of a positive number."[3]

Bhâskara II:

"The square of a positive and a negative number is positive; the square-root of a positive number is positive as well as negative. There is no square-root of a negative number, because it is non-square."[4]

Nârâyaṇa:

"The square of a positive and a negative number is positive. The square-root of a positive number will be positive and also negative. It has been proved that a negative number, being non-square, has no square-root."[5]

[1] *BrSpSi*, xviii. 35.
[2] *GSS*, i. 52.
[3] *SiŚe*, xiv. 5.
[4] *BBi*, p. 4.
[5] *NBi*, I, R. 10.

5. FUNDAMENTAL OPERATIONS

Number of Operations. The number of fundamental operations in algebra is recognised by all Hindu algebraists to be six, *viz.*, addition, subtraction, multiplication, division, squaring and the extraction of the square-root. So the cubing and the extraction of the cube-root which are included amongst the fundamental operations of arithmetic, are excluded from algebra. But the formula

$$(a + b)^3 = a^3 + 3a^2b + 3ab^2 + b^3,$$

or

$$(a + b)^3 = a^3 + 3ab(a + b) + b^3,$$

is found to have been given, as stated before, in almost all the Hindu treatises on arithmetic beginning with that of Brahmagupta (628). By applying it repeatedly, Mahâvîra indicates how to find the cube of an algebraic expression containing more than two terms; thus

$$\begin{aligned}(a + b + c + d + \ldots)^3 &\\ = a^3 + 3a^2(b + c + d + \ldots.) &+ 3a(b + c + d + \ldots)^2 \\ &+ (b + c + d + \ldots.)^3, \\ = a^3 + 3a^2(b + c + d + \ldots.) &+ 3a(b + c + d + \ldots)^2 \\ + b^3 + 3b^2(c + d + \ldots) &+ 3b(c + d + \ldots\ldots)^2 \\ + (c + d + \ldots.)^3; &\end{aligned}$$

and so on.

Addition and Subtraction. Brahmagupta says:

"Of the unknowns, their squares, cubes, fourth powers, fifth powers, sixth powers, etc., addition and subtraction are (performed) of the like; of the unlike (they mean simply their) statement severally."[1]

Bhâskara II:

"Addition and subtraction are performed of those

[1] *BrSpSi*, xviii. 41.

of the same species (*jâti*) amongst unknowns; of different species they mean their separate statement."[1]

Nârâyaṇa:

"Of the colours or letters of alphabets (representing the unknowns) addition should be made of those which are of the same species; and similarly subtraction. In the addition and subtraction of those of different species the result will be their putting down severally."[2]

Multiplication. Brahmagupta says:

"The product of two like unknowns is a square; the product of three or more like unknowns is a power of that designation. The multiplication of unknowns of unlike species is the same as the mutual product of symbols; it is called *bhâvita* (product or factum)."[3]

Bhâskara II writes:

"A known quantity multiplied by an unknown becomes unknown; the product of two, three or more unknowns of like species is its square, cube, etc.; and the product of those of unlike species is their *bhâvita*. Fractions, etc., are (considered) as in the case of knowns; and the rest (*i.e.*, remaining operations) will be the same as explained in arithmetic. The multiplicand is put down separately in as many places as there are terms in the multiplier and is then severally multiplied by those terms; (the products are then) added together according to the methods indicated before. Here, in the squaring and multiplication of unknowns, should be followed the method of multiplication by component parts, as explained in arithmetic."[4]

The same rules are given by Nârâyaṇa.[5] The fol-

[1] *BBi*, p. 7.
[2] *NBi*, I, R. 19.
[3] *BrSpSi*, xviii. 42.
[4] *BBi*, p. 8.
[5] *NBi*, I, R. 21-2.

lowing illustration amongst others, is given by Bhâskara II:

"Tell at once, O learned, (the result) of multiplying five *yâvat-tâvat* minus one known quantity by three *yâvat-tâvat* plus two knowns.

"Statement: Multiplicand *yâ* 5 *rû* $\dot{1}$; multiplier *yâ* 3 *rû* 2; on multiplication the product becomes *yâ va* 15 *yâ* 7 *rû* $\dot{2}$."[1]

The detailed working of this illustration is shown by the commentator Kṛṣṇa (*c.* 1580) thus:

"*yâ* 3	*yâ* 5	*rû* $\dot{1}$	*yâ va* 15	*yâ* $\dot{3}$	
rû 2	*yâ* 5	*rû* $\dot{1}$		*yâ* 10	*rû* $\dot{2}$ "
			yâ va 15	*yâ* 7	*rû* $\dot{2}$

Division. Bhâskara II states:

"By whatever unknowns and knowns, the divisor is multiplied (severally) and subtracted from the dividend successively so that no residue is left, they constitute the quotients at the successive stages."[2]

Nârâyaṇa describes the method of division in nearly the same terms.[3] As an example of division, Bhâskara II proposes to divide $18x^2 + 24xy - 12xz - 12x + 8y^2 - 8yz - 8y + 2z^2 + 4z + 2$ by $-3x - 2y + z + 1$. He simply states the quotient without indicating the different steps in the process. Kṛṣṇa supplies the details of the process which are substantially the same as followed at present.

Squaring. Only one rule for the squaring of an algebraic expression is found in treatises on algebra. It is the same as that stated before in the treatises on arithmetic, *viz.*,

$$(a + b)^2 = a^2 + b^2 + 2ab;$$

[1] *BBi*, p. 8. [2] *BBi*, p. 9.

[3] *NBi*, I, R. 23.

or, in its general form,

$$(a+b+c+d+\ldots)^2=a^2+b^2+c^2+d^2+\ldots+2\sum ab.$$

Square-root. For finding the square-root of an algebraic expression Bhâskara II gives the following rule:

"Find the square-root of the unknown quantities which are squares; then deduct from the remaining terms twice the products of those roots two and two; if there be known terms, proceed with the remainder in the same way after taking the square-root of the knowns."[1]

Nârâyaṇa says:

"First find the root of the square terms (of the given expression); then the product of two and two of them (roots) multiplied by two should be subtracted from the remaining terms; (the result thus obtained) is said to be the square-root here (in algebra)."[2]

Jñânarâja writes:

"Take the square-root of those (terms) which are capable of yielding roots; the product of two and two of these (roots) multiplied by two should be deducted from the remaining terms of that square (expression); the result will be the (required) root, so say the experts in this (science)."

6. EQUATIONS

Forming Equations. Before proceeding to the actual solution of an equation of any type, certain preliminary operations have necessarily to be carried out in order to prepare it for solution. Still more preliminary work is that of forming the equation (*samî-karaṇa*, *samî-kâra* or *samî-kriyâ*; from *sama*, equal and

[1] *BBi*, p. 10. [2] *NBi*, I, R. 24.

kṛ, to do; hence literally, making equal) from the conditions of the proposed problem. Such preliminary work may require the application of one or more fundamental operations of algebra or arithmetic. The operations preliminary to the formation of a simple equation have been described by Pṛthûdakasvâmî (860) thus:

"In this case, in the problem proposed by the questioner, *yâvat-tâvat* is put for the value of the unknown quantity; then performing multiplication, division, etc., as required in the problem the two sides shall be carefully made equal. The equation being formed in this way, then the rule (for its solution) follows."[1]

Bhâskara II's descriptions are fuller: He says :

"Let *yâvat-tâvat* be assumed as the value of the unknown quantity. Then doing precisely as has been specifically told—by subtracting, adding, multiplying or dividing[2]—the two equal sides (of an equation) should be very carefully built."[3]

Nârâyaṇa says:

"Of these (four classes of equations), the linear equation in one unknown (will be treated) first. In a problem (proposed), the value of the quantity which is unknown is assumed to be *yâvat*, one, two or any multiple of it, with or without an absolute term, which again may be additive or subtractive. Then on the value thus assumed optionally should be performed, in

[1] *BrSpSi*, xviii. 43 (*Com*).

[2] In his gloss Bhâskara II explains: "Every operation, such as multiplication, division, rule of three, rule of five terms, summation of series, or treatment of plane figures, etc., according to the statement of the problem should be performed......" See *BBi*, p. 44.

[3] *BBi*, p. 43.

accordance with the statement of the problem, the operations such as addition, subtraction, multiplication, division, rule of three, double rule of three, summation, plane figures, excavations, etc. And thus the two sides must be made equal. If the equality of the two sides is not explicitly stated, then one side should be multiplied, divided, increased or decreased by one's own intelligence (according to the problem) and thus the two sides must be made equal."[1]

Plan of Writing Equations. After an equation is formed, writing it down for further operations is technically called *nyâsa* (putting down, statement) of the equation. In the Bakhshâlî treatise the two sides of an equation are put down one after the other in the same line without any sign of equality being interposed.[2] Thus the equations

$$\sqrt{x+5}=s,\ \sqrt{x-7}=t$$

appear as[3]

0	5	*yu*	*mû*	0	*sa*	0	7	+	*mû*	0
1	1			1		1	1			1

The equation

$$x+2x+3\times 3x+12\times 4x=300$$

is stated as[4]

0	2	1	3	3	12	4	*dṛśya* 300.
1	1	1	1	1	1	1	

This plan of writing an equation was subsequently abandoned by the Hindus for a new one in which the two sides are written one below the other without any

[1] *NBi*, II, R. 3 (*Gloss*).
[2] Datta, *Bakh. Math.*, (*BCMS*, XXI), p. 28.
[3] Folio 59, recto. [4] Folio 23, verso.

sign of equality. Further, in this new plan, the terms of similar denominations are usually written one below the other and even the terms of absent denominations on either side are expressly indicated by putting zeros as their coefficients. Reference to the new plan is found as early as the algebra of Brahmagupta (628).[1] Pṛthûdakasvâmî (860) represented the equation[2]

$$10x - 8 = x^2 + 1$$

as follows:

yâ va 0 *yâ* 10 *rû* $\dot{8}$
yâ va 1 *yâ* 0 *rû* 1

which means, writing x for *yâ*

$$x^2.0 + x.10 - 8$$
$$x^2 + x.0 + 1$$

or $$0x^2 + 10x - 8 = x^2 + 0x + 1.$$

If there be several unknowns, those of the same kind are written in the same column with zero coefficients, if necessary. Thus the equation

$$197x - 1644y - z = 6302$$

is represented by Pṛthûdakasvâmî thus:[3]

yâ 197 *kâ* $16\dot{4}\dot{4}$ *nî* $\dot{1}$ *rû* 0
yâ 0 *kâ* 0 *nî* 0 *rû* 6302

which means, putting y for *kâ* and z for *nî*,

$$197x - 1644y - z + 0 = 0x + 0y + 0z + 6302.$$

The following two instances are from the *Bîjagaṇita* of Bhâskara II (1150):[4]

(1) *yâ* 5 *kâ* 8 *nî* 7 *rû* 90
yâ 7 *kâ* 9 *nî* 6 *rû* 62

[1] *BrSpSi*, xvii. 43 (*vide infra*, p. 33). Compare also *BBi*, p. 127.
[2] *BrSpSi*, xviii. 49 (*Com*).
[3] *BrSpSi*, xviii. 54 (*Com*).
[4] *BBi*, pp. 78, 101.

means (writing x for *yâ*, y for *kâ* and z for *nî*)

$$5x + 8y + 7z + 90 = 7x + 9y + 6z + 62.$$

(2) *yâ gha* 8 *yâ va* 4 *kâ va yâ. bhâ* 10

yâ gha 4 *yâ va* 0 *kâ va yâ. bhâ* 12

means $8x^3 + 4x^2 + 10y^2x = 4x^3 + 0x^2 + 12y^2x,$

or $8x^3 + 4x^2 + 10y^2x = 4x^3 + 12y^2x.$

In the above plan it will be noticed that the terms are ordered according to descending powers of the unknowns. Numerical coefficients are placed after the unknowns; if the coefficient be unity it is noted particularly and if the coefficient be fractional it is kept distinct from the unknown, that is, its denominator is so written as not to come under the unknown;[1] the minus sign is put over the numerical coefficient rather than on the unknown; and the absolute term is invariably put last on either side. As has been observed by Professor Smith,[2] this plan "in one respect was the best that has ever been suggested." For "it shows at a glance the similar terms one above the other, and permits of easy transposition."

The use of the old plan of writing equations is sometimes met with in later works also. For instance, in the MS. of Pṛthûdakasvâmî's commentary on the *Brâhma-sphuṭa-siddhânta*, an incomplete copy of which is preserved in the library of the Asiatic Society of Bengal (No. I B6) we find a statement of equations thus: "first side *yâvargaḥ* 1 *yâvakaḥ* 200 *rû* 0; second side *yâvargaḥ* 0 *yâvakaḥ* 0 *rû* 1500."[3] That is to say,

$$x^2 + 200x + 0 = 0x^2 + 0x + 1500.$$

[1] For instance, see *BBi*, pp. 47 ff.
[2] Smith, *History*, II, pp. 425, 426.
[3] *BrSpSi*, xii. 15 (*Com*).

Preparation of Equations. The operation to be performed on an equation next to its statement (*nyâsa*) is technically known as *sâmaśodhana* (from *sama*, meaning equal or complete, and *śodhana*, clearance; hence literally meaning, equi-clearance or complete clearance[1]) or simply *śodhana*. The nature of this clearance varies according to the kind of the equation. In the case of an equation in one unknown only, whether linear, quadratic or of higher powers, one side of it is cleared of the unknowns of all denominations and the other side of it of the absolute terms, so that the equation is ultimately reduced to one of the form

$$ax^2 + bx = c,$$

where a, b, c may be positive or negative; some of them may be even zero. Thus Brahmagupta observes:

"From which the square of the unknown and the unknown are cleared, the known quantities are cleared (from the side) *below* that."[2]

Pṛthûdakasvâmî explains:

"This rule[3] has been introduced for that case in which the two sides of the equation having been formed in accordance with the statement of the problem, there are present the square and other powers of the unknown together with the (simple) unknown. The absolute terms should be cleared off from the side opposite to that from which are cleared the square (and other powers) of the unknown and the (simple) unknown. When perfect clearance (*samaśodhana*) has

[1] Colebrooke's rendering of the term as "equal subtraction", though not literally inaccurate, is technically so; at least it is not happy.

[2] *BrSpSi*, xviii. 43.

[3] The reference is to Brahmagupta's rule for the solution of a quadratic equation.

3

been thus made..."[1]

Srîpati says:

"From one (side) the square of the unknown and the unknown should be cleared by removing the known quantities; the known quantities (should be cleared) from the side opposite to that."[2]

Similarly Bhâskara II:

"Then the unknown on one side of it (the equation) should be subtracted from the unknown on the other side; so also the square and other powers of the unknown; the known quantities on the other side should be subtracted from the known quantities of another (*i.e.*, the former) side."[3]

Here we give a few illustrations. With reference to the equations from the commentary of Pṛthûdakasvâmî, stated on page 31, the author says:

"Perfect clearance (*samaśodhana*) being made in accordance with the rule, (the equation) will be

yâ va 1 *yâ* $\dot{1}\dot{0}$

rû $\dot{9}$"

i.e., $x^2 - 10x = -9.$

The following illustration is from the *Bîjagaṇita* of Bhâskara II:[4]

"Thus the two sides are

yâ va 4 *yâ* $3\dot{4}$ *rû* 72

yâ va 0 *yâ* 0 *rû* 90

On complete clearance (*samaśodhana*), the residues of the two sides are

[1] *BrSpSi*, xviii. 44 (*Com*).
[2] *SiŚe*, xvi. 17.
[3] *BBi*, p. 44.
[4] *BBi*, p. 63.

yâ va 4 *yâ* 34 *rû* 0
yâ va 0 *yâ* 0 *rû* 18"

i.e., $$4x^2 - 34x = 18.$$

Classification of Equations. The earliest Hindu classification of equations seems to have been according to their degrees, such as simple (technically called *yâvat-tâvat*), quadratic (*varga*), cubic (*ghana*) and biquadratic (*varga-varga*). Reference to it is found in a canonical work of *circa* 300 B. C.[1] But in the absence of further corroborative evidence, we cannot be sure of it. Brahmagupta (628) has classified equations as: (1) equations in one unknown (*eka-varṇa-samîkaraṇa*), (2) equations in several unknowns (*aneka-varṇa-samîkaraṇa*), and (3) equations involving products of unknowns (*bhâvita*). The first class is again divided into two subclasses, *viz*., (*i*) linear equations, and (*ii*) quadratic equations (*avyakta-varga-samîkaraṇa*). Here then we have the beginning of our present method of classifying equations according to their degrees. The method of classification adopted by Pṛthûdakasvâmî (860) is slightly different. His four classes are: (1) linear equations with one unknown, (2) linear equations with more unknowns, (3) equations with one, two or more unknowns in their second and higher powers, and (4) equations involving products of unknowns. As the method of solution of an equation of the third class is based upon the principle of the elimination of the middle term, that class is called by the name *madhyamâharaṇa* (from *madhyama*, "middle", *âharaṇa* "elimination", hence meaning "elimination of the middle term"). For the other classes, the old names given by Brahmagupta have been retained. This method of classification has been followed by subsequent writers.

[1] *Sthânâṅga-sûtra*, Sûtra 747. For further particulars see Datta, *Jaina Math.*, (*BCMS*, XXI), pp. 119ff.

Bhâskara II distinguishes two types in the third class, *viz.*, (*i*) equations in one unknown in its second and higher powers and (*ii*) equations having two or more unknowns in their second and higher powers. According to Kṛṣṇa (1580) equations are primarily of two classes: (1) equations in one unknown and (2) equations in two or more unknowns. The class (1), again, comprises two subclasses: (*i*) simple equations and (*ii*) quadratic and higher equations. The class (2) has three subclasses: (*i*) simultaneous linear equations, (*ii*) equations involving the second and higher powers of unknowns, and (*iii*) equations involving products of unknowns. He then observes that these five classes can be reduced to four by including the second subclasses of classes (1) and (2) into one class as *madhyamâharaṇa*.

7. LINEAR EQUATIONS IN ONE UNKNOWN

Early Solutions. As already stated, the geometrical solution of a linear equation in one unknown is found in the *Sulba*, the earliest of which is not later than 800 B.C. There is a reference in the *Sthânâṅga-sûtra* (*c.* 300 B.C.) to a linear equation by its name (*yâvat-tâvat*) which is suggestive of the method of solution[1] followed at that time. We have, however, no further evidence about it. The earliest Hindu record of doubtless value of problems involving simple algebraic equations and of a method for their solution occurs in the Bakhshâlî treatise, which was written very probably about the beginning of the Christian Era. One problem is:[2]

"The amount given to the first is not known. The second is given twice as much as the first; the third

[1] Datta, *Jaina Math.*, (*BCMS*, XXI), p. 122.
[2] *BMs*, Folio 23, recto.

thrice as much as the second; and the fourth four times as much as the third. The total amount distributed is 132. What is the amount of the first?"

If x be the amount given to the first, then according to the probelm,

$$x + 2x + 6x + 24x = 132.$$

Rule of False Position. The solution of this equation is given as follows:

" 'Putting any desired quantity in the vacant place'; any desired quantity is || 1 ||; 'then construct the series'

1	2	2 3	6 4
1	1	1 1	1 1

'multiplied' || 1 | 2 | 6 | 24 |; 'added' 33. 'Divide the visible quantity' $\left|\begin{matrix}132\\33\end{matrix}\right|$; (which) on reduction becomes

$\left|\begin{matrix}4\\1\end{matrix}\right|$. (This is) the amount given (to the first)."[1]

The solution of another set of problems in the Bakhshâlî treatise, leads ultimately to an equation of the type[2]

$$ax + b = p.$$

The method given for its solution is to put any arbitrary value g for x, so that

$$ag + b = p', \text{ say}.$$

Then the correct value will be

$$x = \frac{p - p'}{a} + g.$$

[1] *Ibid.*

[2] *Vide infra*, pp. 48f.

The above method of solution of a linear equation was known in the middle ages, amongst Arab and European algebraists, by the name of the Rule of False Position. It is noteworthy that the terms *yaddṛcchâ*, *vâñcchâ*, and *kâmika* of the Bakhshâlî treatise are equivalent to the term *yâvat-tâvat*. So the origin of this latter term seems to be connected with the Rule of False Position. It is interesting to find that the rule was once given so much importance in Hindu algebra that the section dealing with it was named after it.

Disappearance from Later Algebra. The Rule of False Position bespeaks of an early stage of development of the science of algebra when there was no symbol for the unknown. It naturally disappears with the introduction of a system of notations.[1] This will account for the complete disappearance of the Rule of False Position from the later Hindu treatises on algebra beginning with that of Âryabhaṭa I (499). There are found, however, very limited applications of it in the arithmetical treatises of Śrîdhara (*c.* 750), Mahâvîra (850) and Bhâskara II (1150). This can be accounted for easily. The problems which have been solved by those writers with the help of the Rule of False Position are really of algebraic nature, though incorporated into arithmetical treatises. But as the use of algebraic symbols and notations is not permissible in arithmetic, recourse had to be taken to that Rule. For instance, we take the following problem from the *Gaṇita-sâra-saṁgraha* of Mahâvîra:

"The sum of $\frac{1}{8}$, $\frac{1}{4}$ of $\frac{1}{3}$, $\frac{1}{5}$ of $\frac{1}{2}$, $\frac{1}{5}$ of $\frac{3}{4}$ of $\frac{1}{6}$, of a certain number is equal to $\frac{1}{2}$. What is that unknown (number)?"[2]

Mahâvîra gives the following rule for finding out

[1] Smith, *History*, II, p. 437. [2] *GSS*, iii. 108.

the unknown in a problem of this kind :

"Put down one for the value of the unknown ; then in accordance with the previous rule (find) the sum (of its parts) ; divide the known (number) by that (sum) ; the quotient will be (the value of) the unknown in compound fractions."[1]

Operation with an Optional Number. Bhâskara II describes a method called *Iṣṭa-karma* or "operation with an optional number." This may be illustrated by the following example :

"What is that number which multiplied by five, diminished by its third part and (then) divided by ten, will become, together with its one-third, half and one-fourth parts, equal to seventy minus two ?"[2]

i.e., $$\frac{5x - 5x/3}{10} + \frac{x}{3} + \frac{x}{2} + \frac{x}{4} = 70 - 2.$$

Bhâskara assumes $x = 3$ and then calculates

$$\frac{5 \times 3 - 5 \times 3/3}{10} + \tfrac{3}{3} + \tfrac{3}{2} + \tfrac{3}{4} = \tfrac{17}{4}.$$

He then states

$$x = 68 \times 3 \div \tfrac{17}{4} = 48.$$

He observes : "Similarly, in every example, by whatever the (required) number is multiplied or divided, by whatever fraction of the number it is found to have been increased or diminished, assuming an optional number, on it perform the same operations in accordance with the statement of the problem; by that, which results, divide the known number multiplied by the assumed number; the quotient will be the (required) number."[3]

[1] *GSS*, iii. 107. [2] *L*, p. 10.
[3] *L*, p. 11.

Solution of Linear Equations. Âryabhaṭa I (499) says:

"The difference of the known "amounts" relating to the two persons should be divided by the difference of the coefficients of the unknown.[1] The quotient will be the value of the unknown, if their possessions be equal."[2]

This rule contemplates a problem of this kind: Two persons, who are equally rich, possess respectively a, b times a certain unknown amount together with c, d units of money in cash. What is that amount?

If x be the unknown amount, then by the problem

$$ax + c = bx + d.$$

Therefore $$x = \frac{d - c}{a - b}.$$

Hence the rule.

Brahmagupta says:

"In a (linear) equation in one unknown, the difference of the known terms taken in the reverse order, divided by the difference of the coefficients of the unknown (is the value of the unknown)."[3]

Śrîpati writes:

"First remove the unknown from any one of the sides (of the equation) leaving the known term; the reverse (should be done) on the other side. The difference of the absolute terms taken in the reverse order

[1] The original is *gulikântara* which literally means "the difference of the unknowns." But what is implied is "the difference of the coefficients of the unknown." As has been observed by Pṛthûdakasvâmî, according to the usual practice of Hindu algebra, "the coefficient of the square of the unknown is called the square (of the unknown) and the coefficient of the (simple) unknown is called the unknown." *BrSpSi*, xviii. 44 (*Com*).

[2] *Â*, ii. 30.

[3] *BrSpSi*, xviii. 43.

divided by the difference of the coefficients of the unknown will be the value of the unknown."[1]

Bhâskara II states:

"Subtract the unknown on one side from that on the other and the absolute term on the second from that on the first side. The residual absolute number should be divided by the residual coefficient of the unknown; thus the value of the unknown becomes known."[2]

Nârâyaṇa writes:

"From one side clear off the unknown and from the other the known quantities; then divide the residual known by the residual coefficient of the unknown. Thus will certainly become known the value of the unknown."[3]

For illustration we take a problem proposed by Brahmagupta:

"Tell the number of elapsed days for the time when four times the twelfth part of the residual degrees increased by one, plus eight will be equal to the residual degrees plus one."[4]

It has been solved by Pṛthûdakasvâmî as follows:

"Here the residual degrees are (put as) *yâvat-tâvat*, *yâ*; increased by one, *yâ* 1 *rû* 1; twelfth part of it, $\frac{\textit{yâ}\ 1\ \ \textit{rû}\ 1}{12}$; four times this, $\frac{\textit{yâ}\ 1\ \ \textit{rû}\ 1}{3}$; plus the absolute quantity eight, $\frac{\textit{yâ}\ 1\ \ \textit{rû}\ 25}{3}$. This is equal to the residual degrees plus unity. The statement of both sides tripled is

yâ 1 *rû* 25
yâ 3 *rû* 3

[1] *SiŚe*, xiv. 15.
[2] *BBi*, p. 44.
[3] *NBi*, II, R. 5.
[4] *BrSpSi*, xviii. 46.

The difference between the coefficients of the unknown is 2. By this the difference of the absolute terms, namely 22, being divided, is produced the residual of the degrees of the sun, 11. These residual degrees should be known to be irreducible. The elapsed days can be deduced then, (proceeding) as before."

In other words, we have to solve the equation

$$\tfrac{4}{12}(x + 1) + 8 = x + 1,$$

which gives $$x + 25 = 3x + 3,$$

or $$2x = 22.$$

Therefore $$x = 11.$$

The following problem and its solution are from the *Bîjagaṇita* of Bhâskara II :

"One person has three hundred coins and six horses. Another has ten horses (each) of similar value and he has further a debt of hundred coins. But they are of equal worth. What is the price of a horse ?

"Here the statement for equi-clearance is :

$$6x + 300 = 10x - 100.$$

Now, by the rule, 'Subtract the unknown on one side from that on the other etc.,' unknown on the first side being subtracted from the unknown on the other side, the remainder is $4x$. The absolute term on the second side being subtracted from the absolute term on the first side, the remainder is 400. The residual known number 400 being divided by the coefficient of the residual unknown $4x$, the quotient is recognised to be the value of x, (namely) 100."[1]

There are a few instances in the Bakhshâlî work where a method similar to that of later writers appears

[1] *BBi*, pp. 44f.

to have been followed for the solution of a linear equation. One such problem is: Two persons start with different initial velocities (a_1, a_2); travel on successive days distances increasing at different rates (b_1, b_2). But they cover the same distance after the same period of time. What is the period?

Denoting the period by x, we get

$$a_1 + (a_1 + b_1) + (a_1 + 2b_1) + \ldots \text{ to } x \text{ terms}$$
$$= a_2 + (a_2 + b_2) + (a_2 + 2b_2) + \ldots \text{ to } x \text{ terms},$$

or
$$\left\{a_1 + \left(\frac{x-1}{2}\right)b_1\right\}x = \left\{a_2 + \left(\frac{x-1}{2}\right)b_2\right\}x;$$

whence
$$x = \frac{2(a_2 - a_1)}{b_1 - b_2} + 1,$$

which is the solution given in the Bakhshâlî work:

"Twice the difference of the initial terms divided by the difference of the common differences, is increased by unity. The result will be the number of days in which the distance moved will be the same."[1]

8. LINEAR EQUATIONS WITH TWO UNKNOWNS

Rule of Concurrence. One topic commonly discussed by almost all Hindu writers goes by the special name of *saṅkramaṇa* (concurrence). According to Nârâyaṇa (1350), it is also called *saṅkrama* and *saṅkrâma*.[2] Brahmagupta (628) includes it in algebra while others consider it as falling within the scope of arithmetic. As explained by the commentator Gaṅgâdhara (1420), the subject of discussion here is "the investigation of two quantities concurrent or grown together in the form of their sum and difference."

[1] *BMs*, Folio 4, verso. [2] *GK*, i. 31.

Or, in other words, *saṅkramaṇa* is the solution of the simultaneous equations

$$x + y = a,$$
$$x - y = b.$$

So Brahmagupta and Śrîpati are perfectly right in thinking that concurrence is truly a topic for algebra.

Brahmagupta's rule for solution is :

"The sum is increased and diminished by the difference and divided by two ; (the result will be the two unknown quantities) : (this is) concurrence."[1]

The same rule is restated by him on a different occasion in the form of a problem and its solution.

"The sum and difference of the residues of two (heavenly bodies) are known in degrees and minutes. What are the residues ? The difference is both added to and subtracted from the sum, and halved ; (the results are) the residues."[2]

Similar rules are given also by other writers.[3]

Linear Equations. Mahâvîra gives the following examples leading to simultaneous linear equations together with rules for the solution of each.

Example. "The price of 9 citrons and 7 fragrant wood-apples taken together is 107 ; again the price of 7 citrons and 9 fragrant wood-apples taken together is 101. O mathematician, tell me quickly the price of a citron and of a fragrant wood-apple quite separately."[4]

If x, y be the prices of a citron and of a fragrant

[1] *BrSpSi*, xviii. 36. [2] *BrSpSi*, xviii. 96.
[3] *GSS*, vi. 2 ; *MSi*, xv. 21 ; *SiŚe*, xiv. 13 ; *L*, p. 12 ; *GK*, i. 31.
[4] *GSS*, vi. 140½–142½.

wood-apple respectively, then

$$9x + 7y = 107,$$
$$7x + 9y = 101.$$

Or, in general,

$$ax + by = m,$$
$$bx + ay = n.$$

Solution. "From the larger amount of price multiplied by the (corresponding) bigger number of things subtract the smaller amount of price multiplied by the (corresponding) smaller number of things. (The remainder) divided by the difference of the squares of the numbers of things will be the price of each of the bigger number of things. The price of the other will be obtained by reversing the multipliers."[1]

Thus $x = \dfrac{am - bn}{a^2 - b^2}, \quad y = \dfrac{an - bm}{a^2 - b^2}.$

Example. "A wizard having powers of mystic incantations and magical medicines seeing a cock-fight going on, spoke privately to both the owners of the cocks. To one he said, 'If your bird wins, then you give me your stake-money, but if you do not win, I shall give you two-thirds of that.' Going to the other, he promised in the same way to give three-fourths. From both of them his gain would be only 12 gold pieces. Tell me, O ornament of the first-rate mathematicians, the stake-money of each of the cock-owners."[2]

i.e., $x - \frac{3}{4}y = 12, \quad y - \frac{2}{3}x = 12.$

Or, in general,

$$x - \frac{c}{d}y = p, \quad y - \frac{a}{b}x = p.$$

[1] *GSS*, vi. $139\frac{1}{2}$.

[2] *GSS*, vi. 270-$2\frac{1}{2}$.

Solution:[1]

$$x = \frac{b(c+d)}{(c+d)b-(a+b)c}p,$$

$$y = \frac{d(a+b)}{(a+b)d-(c+d)a}p.$$

The following example with its solution is taken from the *Bījagaṇita* of Bhâskara II :

Example. "One says, 'Give me a hundred, friend, I shall then become twice as rich as you.' The other replies, 'If you give me ten, I shall be six times as rich as you.' Tell me what is the amount of their (respective) capitals ?"[2]

The equations are

$$x + 100 = 2(y - 100), \qquad (1)$$

$$y + 10 = 6(x - 10). \qquad (2)$$

Bhâskara II indicates two methods of solving these equations. They are substantially as follows :

First Method.[3] Assume

$$x = 2z - 100, \quad y = z + 100,$$

so that equation (1) is identically satisfied. Substituting these values in the other equation, we get

$$z + 110 = 12z - 660;$$

whence $z = 70$. Therefore, $x = 40$, $y = 170$.

Second Method.[4] From equation (1), we get

$$x = 2y - 300,$$

and from equation (2)

$$x = \tfrac{1}{6}(y + 70).$$

[1] *GSS*, vi. 268½-9½.
[2] *BBi*, p. 41.
[3] *BBi*, p. 46.
[4] *BBi*, pp. 78f.

Equating these two values of x, we have

$$2y - 300 = \tfrac{1}{6}(y + 70),$$

or
$$12y - 1800 = y + 70;$$

whence $y = 170$. Substituting this value of y in any of the two expressions for x, we get $x = 40$.

It is noteworthy that the second method of solution of the problem under consideration is described by Bhâskara II in the section of his algebra dealing with "linear equations with several unknowns," while the first method in that dealing with "linear equations in one unknown." In this latter connection he has observed that the solution of a problem containing two unknowns can sometimes be made by ingenious artifices to depend upon the solution of a simple linear equation.

9. LINEAR EQUATIONS WITH SEVERAL UNKNOWNS

A Type of Linear Equations. The earliest Hindu treatment of systems of linear equations involving several unknowns is found in the Bakhshâlî treatise. One problem in it runs as follows :

"[Three persons possess a certain amount of riches each.] The riches of the first and the second taken together amount to 13 ; the riches of the second and the third taken together are 14 ; and the riches of the first and the third mixed are known to be 15. Tell me the riches of each."[1]

If x_1, x_2, x_3 be the wealths of the three merchants respectively, then

$$x_1 + x_2 = 13,\ x_2 + x_3 = 14,\ x_3 + x_1 = 15. \qquad (1)$$

Another problem is

[1] *BMs*, Folio 29, recto. The portions within [ ] in this and the following illustration have been restored.

"[Five persons possess a certain amount of riches each. The riches of the first] and the second mixed together amount to 16; the riches of the second and the third taken together are known to be 17; the riches of the third and the fourth taken together are known to be 18; the riches of the fourth and the fifth mixed together are 19; and the riches of the first and the fifth together amount to 20. Tell me what is the amount of each."[1]

i.e., $x_1 + x_2 = 16,\ x_2 + x_3 = 17,\ x_3 + x_4 = 18,$

$$x_4 + x_5 = 19,\ x_5 + x_1 = 20. \qquad (2)$$

There are in the work a few other similar problems.[2] Every one of them belongs to a system of linear equations of the type

$$x_1 + x_2 = a_1,\ x_2 + x_3 = a_2,\ \ldots,\ x_n + x_1 = a_n, \qquad \text{(I)}$$

n being *odd*.

Solution by False Position. A system of linear equations of this type is solved in the Bakhshâlî treatise substantially as follows:

Assume an arbitrary value p for x_1 and then calculate the values of x_2, x_3, ... corresponding to it. Finally let the calculated value of $x_n + x_1$ be equal to b (say). Then the true value of x_1 is obtained by the formula

$$x_1 = p + \tfrac{1}{2}(a_n - b).$$

In the particular case (1) the author[3] assumes the arbitrary value 5 for x_1; then are successively calculated the values $x'_2 = 8$, $x'_3 = 6$ and $x'_3 + x'_1 = 11$. The correct values are, therefore,

$$x_1 = 5 + (15 - 11)/2 = 7,\ x_2 = 6,\ x_3 = 8.$$

[1] *BMs*, Folios 27 and 29, verso.
[2] *BMs*, Folio 30, recto; also see Kaye's Introduction, p. 40.
[3] *BMs*, Folio, 29, recto.

Rationale. By the process of elimination we get from equations (I)

$$(a_2-a_1)+(a_4-a_3)+\ldots+(a_{n-1}-a_{n-2})+2x_1=a_n.$$

Assume $x_1=p$; so that

$$(a_2-a_1)+(a_4-a_3)+\ldots\ldots+(a_{n-1}-a_{n-2})+2p=b, \text{ say.}$$

Subtracting $\quad 2(x_1-p)=a_n-b.$

Therefore $\quad x_1=p+\frac{1}{2}(a_n-b).$

Second Type. A particular case of the type of equations (I) for which $n=3$, may also be looked upon as belonging to a different type of systems of linear equations, *viz.*,

$$\Sigma x-x_1=a_1, \Sigma x-x_2=a_2, \ldots, \Sigma x-x_n=a_n, \quad \text{(II)}$$

where Σx stands for $x_1+x_2+\ldots+x_n$. But it will not be proper to say that equations of this type have been treated in the Bakhshâlî treatise.[1] They have, however, been solved by Âryabhaṭa (499) and Mahâvîra (850). The former says:

"The (given) sums of certain (unknown) numbers, leaving out one number in succession, are added together separately and divided by the number of terms less one; that (quotient) will be the value of the whole."[2]

i.e., $\quad \Sigma x=\sum_{r=1}^{n} a_r/(n-1).$

Mahâvîra states the solution thus:

"The stated amounts of the commodities added together should be divided by the number of men less

[1] The example cited by Kaye (*BMs*, Introd., p. 40, Ex. vi) which conforms to this type of equations is based upon a misapprehension of the text.

[2] *Â*, ii. 29.

4

one. The quotient will be the total value (of all the commodities). Each of the stated amounts being subtracted from that, (the value) in the hands (of each will be found)."[1]

In formulating his rule Mahâvîra had in view the following example:

"Four merchants were each asked separately by the customs officer about the total value of their commodities. The first merchant, leaving out his own investment, stated it to be 22; the second stated it to be 23, the third 24 and the fourth 27; each of them deducted his own amount in the investment. O friend, tell me separately the value of (the share of) the commodity owned by each."[2]

$$\text{Here}\quad x_1 + x_2 + x_3 + x_4 = \frac{22+23+24+27}{4-1} = 32.$$

$$\text{Therefore}\quad x_1 = 10, \quad x_2 = 9, \quad x_3 = 8, \quad x_4 = 5.$$

Nârâyaṇa says:

"The sum of the depleted amounts divided by the number of persons less one, is the total amount. On subtracting from it the stated amounts severally will be found the different amounts."[3]

The above type of equations is supposed by some modern historians of mathematics[4] to be a modification of the type considered by the Greek Thymaridas and solved by his well known rule *Epanthema*, namely,[5]

$$x + x_1 + x_2 + \ldots + x_{n-1} = s,$$
$$x + x_1 = a_1, \; x + x_2 = a_2, \ldots, x + x_{n-1} = a_{n-1}.$$

[1] *GSS*, vi. 159. [2] *GSS*, vi. 160-2.

[3] *GK*, ii. 28.

[4] Cantor, *Vorlesungen über Geschichte der Mathematik* (referred to hereafter as Cantor, *Geschichte*), I, p. 624; Kaye, *Ind. Math.*, p. 13; *JASB*, 1908, p. 135.

[5] Heath, *Greek Math.*, I, p. 94.

The solution given is

$$x = \frac{(a_1 + a_2 + \dots + a_{n-1}) - s}{n - 2}.$$

But that supposition has been disputed by others.[1] Sarada Kanta Ganguly has shown that it is based upon a misapprehension. It will be noticed that in the Thymaridas type of linear equations, the value of the sum of the unknowns is given whereas in the Āryabhaṭa type it is not known. In fact, Āryabhaṭa determines only that value.

Third Type. A more generalised system of linear equations will be

$$b_1 \Sigma x - c_1 x_1 = a_1, \quad b_2 \Sigma x - c_2 x_2 = a_2, \dots,$$
$$b_n \Sigma x - c_n x_n = a_n. \qquad \text{(III)}$$

Therefore $$\Sigma x = \frac{\Sigma(a/c)}{\Sigma(b/c) - 1}.$$

Hence $$x_r = \frac{b_r}{c_r} \cdot \frac{\Sigma(a/c)}{\Sigma(b/c) - 1} - \frac{a_r}{c_r}, \qquad (1)$$

$$r = 1, 2, 3, \dots, n.$$

A particular case of this type is furnished by the following example of Mahâvîra:

"Three merchants begged money mutually from one another. The first on begging 4 from the second and 5 from the third became twice as rich as the others. The second on having 4 from the first and 6 from the third became thrice as rich. The third man on begging 5 from the first and 6 from the second became five times as rich as the others. O mathematician, if you know

[1] Rodet, *Leçons de Calcul d'Âryabhaṭa*, *JA*, XIII (7), 1878; Sarada Kanta Ganguly, "Notes on Aryabhaṭa," *Jour. Bihar and Orissa Research Soc.*, XII, 1926, pp. 88ff.

the *citra-kuṭṭaka-miśra*,[1] tell me quickly what was the amount in the hand of each."[2]

That is, we get the equations

$$x + 4 + 5 = 2(y + z - 4 - 5),$$
$$y + 4 + 6 = 3(z + x - 4 - 6),$$
$$z + 5 + 6 = 5(x + y - 5 - 6);$$

or

$$2(x + y + z) - 3x = 27,$$
$$3(x + y + z) - 4y = 40,$$
$$5(x + y + z) - 6z = 66;$$

a particular case of the system (III). Substituting in (1), we get

$$x = 7, \quad y = 8, \quad z = 9.$$

In general, suppose $a_{r,1}, a_{r,2}, \ldots a_{r,r-1}, a_{r,r+1} \ldots a_{r,n}$ to be the amounts begged by the rth merchant from the others; and x_r the amount that he had initially. Then

$$x_1 + \Sigma' a_{1,m} = b_1(\Sigma x - x_1 - \Sigma' a_{1,m}),$$
$$x_2 + \Sigma' a_{2,m} = b_2(\Sigma x - x_2 - \Sigma' a_{2,m}),$$
$$\ldots\ldots\ldots\ldots\ldots\ldots\ldots$$
$$x_n + \Sigma' a_{n,m} = b_n(\Sigma x - x_n - \Sigma' a_{n,m}),$$

where $\Sigma' a_{r,m}$ denotes summation from $m = 1$ to $m = n$ excluding $m = r$. Therefore

$$\Sigma x + (b_1 + 1)\Sigma' a_{1,m} = (b_1 + 1)(\Sigma x - x_1),$$
$$\Sigma x + (b_2 + 1)\Sigma' a_{2,m} = (b_2 + 1)(\Sigma x - x_2),$$
$$\ldots\ldots\ldots\ldots\ldots\ldots\ldots$$
$$\Sigma x + (b_n + 1)\Sigma' a_{n,m} = (b_n + 1)(\Sigma x - x_n).$$

Let $\qquad k_r = (b_r + 1)\Sigma' a_{r,m}, \quad r = 1, 2, 3, \ldots, n.$

[1] This is the name given by Mahâvîra to problems involving equations of type (III).

[2] *GSS*, vi. 253½-5½.

Then dividing the foregoing equations by b_1+1, $b_2+1, \ldots$, respectively, and adding together, we get

$$\Sigma x\left(\frac{1}{b_1+1}+\frac{1}{b_2+1}+\ldots+\frac{1}{b_n+1}\right)$$
$$+\left(\frac{k_1}{b_1+1}+\frac{k_2}{b_2+1}+\ldots+\frac{k_n}{b_n+1}\right)=(n-1)\Sigma x.$$

$$\therefore \quad \Sigma x=\left(\frac{k_1}{b_1+1}+\frac{k_2}{b2+1}+\ldots+\frac{k_n}{b_n+1}\right)$$
$$\div\left(\frac{b_1}{b_1+1}+\frac{b_2}{b_2+1}+\ldots+\frac{b_n}{b_n+1}-1\right).$$

Whence

$$x_r=\left\{\frac{k_r+b_rk_1}{b_1+1}+\frac{k_r+b_rk_2}{b_2+1}+\ldots+\frac{k_r+b_rk_{r-1}}{b_{r-1}+1}\right.$$
$$\left.+\frac{k_r+b_rk_{r+1}}{b_{r+1}+1}+\ldots+\frac{k_r+b_rk_n}{b_n+1}-(n-2)k_r\right\}$$
$$\div(b_r+1)\left(\frac{b_1}{b_1+1}+\frac{b_2}{b_2+1}+\ldots+\frac{b_n}{b_n+1}-1\right).$$

Mahâvîra describes the solution thus:

"The sum of the amounts begged by each person is multiplied by the multiple number relating to him as increased by unity. With these (products), the amounts at hand are determined according to the rule *Iṣṭaguṇaghna* etc.[1] They are reduced to a common denominator, and then divided by the sum diminished by unity of the multiple numbers divided by themselves as increased by unity. (The quotients) should be known to be the amounts in the hands of the persons."[2]

Problems of the above kind have been treated also by Nârâyaṇa (1357). He says:

[1] The reference is to rule vi. 241.

[2] *GSS*, vi. 251½-252½.

"Multiply the sum of the monies received by each person by his multiple number plus unity. Then proceed as in the method for "the purse of discord." Divide the multiple number related to each by the same as increased by unity. By the sum diminished by unity of these quotients, divide the results just obtained. The quotients will be the several amounts in their possession."[1]

One particular case, where $b_1 = b_2 = \ldots b_n = 1$ and $c_1 = c_2 = \ldots = c_n = c$, was treated by Brahmagupta (628). He gave the rule:

"The total value (of the unknown quantities) plus or minus the individual values (of the unknowns) multiplied by an optional number being severally (given), the sum (of the given quantities) divided by the number of unknowns increased or decreased by the multiplier will be the total value; thence the rest (can be determined)."[2]

$$\Sigma x \pm cx_1 = a_1,\ \Sigma x \pm cx_2 = a_2, \ldots, \Sigma x \pm cx_n = a_n.$$

Therefore $$\Sigma x = \frac{a_1 + a_2 + \ldots + a_n}{n \pm c}.$$

Hence $$x_1 = \frac{1}{c}\left(\pm a_1 \mp \frac{a_1 + a_2 + \ldots + a_n}{n \pm c}\right);$$

and so on.

Brahmagupta's Rule. Brahmagupta (628) states the following rule for solving linear equations involving several unknowns:

"Removing the other unknowns from (the side of) the first unknown and dividing by the coefficient of the first unknown, the value of the first unknown (is obtained). In the case of more (values of the first unknown),

[1] *GK*, ii. 33f. [2] *BrSpSi*, xiii. 47.

two and two (of them) should be considered after reducing them to common denominators. And (so on) repeatedly. If more unknowns remain (in the final equation), the method of the pulveriser (should be employed). (Then proceeding) reversely (the values of other unknowns can be found)."[1]

Pṛthûdakasvâmî (860) has explained it thus:

"In an example in which there are two or more unknown quantities, colours such as *yâvat-tâvat*, etc., should be assumed for their values. Upon them should be performed all operations conformably to the statement of the example and thus should be carefully framed two or more sides and also equations. Equi-clearance should be made first between two and two of them and so on to the last: from one side one unknown should be cleared, other unknowns reduced to a common denominator and also the absolute numbers should be cleared from the side opposite. The residue of other unknowns being divided by the residual coefficient of the first unknown will give the value of the first unknown. If there be obtained several such values, then with two and two of them, equations should be formed after reduction to common denominators. Proceeding in this way to the end find out the value of one unknown. If that value be (in terms of) another unknown then the coefficients of those two will be reciprocally the values of the two unknowns. If, however, there be present more unknowns in that value, the method of the pulveriser should be employed. Arbitrary values may then be assumed for some of the unknowns."

It will be noted that the above rule embraces the indeterminate as well as the determinate equations. In fact, all the examples given by Brahmagupta in illustra-

[1] *BrSpSi*, xviii. 51.

tion of the rule are of indeterminate character. We shall mention some of them subsequently at their proper places. So far as the determinate simultaneous equations are concerned, Brahmagupta's method for solving them will be easily recognised to be the same as our present one.

Mahâvîra's Rules. Certain interest problems treated by Mahâvîra lead to simple simultaneous equations involving several unknowns. In these problems certain capital amounts ($c_1, c_2, c_3, \ldots$) are stated to have been lent out at the same rate of interest (r) for different periods of time ($t_1, t_2, t_3, \ldots$). If ($i_1, i_2, i_3, \ldots$) be the interests accrued on the several capitals,

$$i_1 = \frac{rt_1c_1}{100},\ i_2 = \frac{rt_2c_2}{100},\ i_3 = \frac{rt_3c_3}{100}, \ldots .$$

(i) If $i_1 + i_2 + i_3 + \ldots . = I$, c_r and t_r be known (for $r = 1, 2, \ldots$), we have

$$i_1 = \frac{Ic_1t_1}{c_1t_1 + c_2t_2 + c_3t_3 + \ldots .},$$

with similar values for $i_2, i_3, \ldots .$

(ii) Or, if $c_1 + c_2 + c_3 + \ldots = C$, i_r and t_r be known (for $r = 1, 2, \ldots$), we have

$$c_1 = \frac{Ci_1/t_1}{i_1/t_1 + i_2/t_2 + \ldots . .},$$

and so on.

(iii) Or, if we are given the sum of the periods $t_1 + t_2 + \ldots = T$, c_r and i_r, then

$$t_1 = \frac{Ti_1/c_1}{i_1/c_1 + i_2/c_2 + \ldots . .},$$

with similar values for $t_2, t_3, \ldots .$

Mahâvîra has given separate rules for the solution

of problems of each of the above three kinds.[1]

Bhâskara's Rule. Bhâskara II has given practically the same rule as that of Brahmagupta for the solution of simultaneous linear equations involving several unknowns. We take the following illustrations from his works.

Example 1. "Eight rubies, ten emeralds and a hundred pearls which are in thy ear-ring were purchased by me for thee at an equal amount; the sum of the price rates of the three sorts of gems is three less than the half of a hundred. Tell me, O dear auspicious lady, if thou be skilled in mathematics, the price of each."[2]

If x, y, z be the prices of a ruby, emerald and pearl respectively, then

$$8x = 10y = 100z,$$

$$x + y + z = 47.$$

Assuming the equal amount to be w, says Bhâskara II, we shall get

$$x = w/8, \quad y = w/10, \quad z = w/100.$$

Substituting in the remaining equation, we easily get $w = 200$. Therefore

$$x = 25, \quad y = 20, \quad z = 2.$$

Example 2. "Tell the three numbers which become equal when added with their half, one-fifth and one-ninth parts, and each of which, when diminished by those parts of the other two, leaves sixty as remainder."[3]

Here we have the equations

$$x + x/2 = y + y/5 = z + z/9, \tag{1}$$

$$x - \frac{y}{5} - \frac{z}{9} = y - \frac{z}{9} - \frac{x}{2} = z - \frac{x}{2} - \frac{y}{5} = 60. \tag{2}$$

[1] *GSS*, vi. 37, 39, 42.

[2] *BBi*, p. 47.

[3] *BBi*, p. 52.

Bhâskara puts w for each of the equal quantities in (1), so that

$$x = \tfrac{2}{3}w,\ y = \tfrac{5}{6}w,\ z = \tfrac{9}{10}w.$$

Substituting these values in (2), any one of them will give

$$\tfrac{2}{5}w = 60;$$

whence $w = 150$. Therefore

$$x = 100,\ y = 125,\ z = 135.$$

It should be noted that the equations (2) are sufficient for the determination of the unknowns.

Example 3. Another type of problem is: Three portions (x, y, z) of a sum of money (c) were lent out at three different rates of interest (r_1, r_2, r_3 per cent per month) for three different periods (t_1, t_2, t_3 months). The interests accrued on them severally were the same. What were those portions ?[1]

i.e.,
$$x + y + z = c, \tag{1}$$

$$\frac{xr_1t_1}{100} = \frac{yr_2t_2}{100} = \frac{zr_3t_3}{100} = I. \tag{2}$$

From (2)

$$x = \frac{100\,I}{r_1t_1},\quad y = \frac{100\,I}{r_2t_2},\quad z = \frac{100\,I}{r_3t_3}.$$

Substituting in (1), we get

$$100\,I\left(\frac{1}{r_1t_1} + \frac{1}{r_2t_2} + \frac{1}{r_3t_3}\right) = c.$$

Therefore
$$I = \frac{c}{100\left(\frac{1}{r_1t_1} + \frac{1}{r_2t_2} + \frac{1}{r_3t_3}\right)}.$$

[1] *L*, p. 22.

Hence
$$x = \frac{\frac{100\times 1}{r_1t_1}\times c}{\frac{100\times 1}{r_1t_1}+\frac{100\times 1}{r_2t_2}+\frac{100\times 1}{r_3t_3}},$$
with similar values for y and z. Bhâskara II says:

"The arguments multiplied by their respective times are divided by the fruit taken into elapsed times. They being divided by their sum and multiplied by the total amount give the portions severally lent out."[1]

10. QUADRATIC EQUATIONS

Early Treatment. It has been shown before that the altar-construction of the Vedic Hindus involved the solution of the complete quadratic equation

$$ax^2 + bx = c, \qquad \text{(I)}$$

as well as of the pure quadratic $ax^2 = c$. The equation that had to be solved was

$$7x^2 + \tfrac{1}{2}x = 7\tfrac{1}{2} + m,$$

which gives $\quad x = \frac{1}{28}(\sqrt{841 + 112m} - 1),$

or $\quad x^2 = \frac{1}{784}\{842 + 112m - 2\sqrt{841 + 112m}\}.$

Simplifying and neglecting higher powers of m, we get

$$x^2 = 1 + \frac{4m}{29}, \text{ approximately.}$$

Kâtyâyana gives the value[2]

$$x^2 = 1 + \frac{m}{7}.$$

[1] *L*, pp. 21f. See also *GT*, 123. If the rate of interest be given as r per p per t months, then p is the argument, t the time and r the fruit. *Cf.* Part I, p. 204; also pp. 225-226.

[2] Datta, *Śulba*, p. 167.

The geometrical solution of the simple quadratic equation

$$4h^2 - 4dh = -c^2$$

is found in the early canonical works of the Jainas (500-300 B. C.) and also in the *Tattvârthâdhigama-sûtra* of Umâsvâti (*c.* 150 B. C.),[1] as

$$h = \tfrac{1}{2}(d - \sqrt{d^2 - c^2}).$$

Bakhshâlî Treatise. The solution of the quadratic equation was certainly known to the author of the Bakhshâlî treatise (*c.* 200). In this work there are some problems of the following type: A certain person travels *s* *yojana* on the first day and *b* *yojana* more on each successive day. Another who travels at the uniform rate of *S* *yojana* per day, has a start of *t* days. When will the first man overtake the second?

If x be the number of days after which the first overtakes the second, then we shall have

$$S(t + x) = x\left\{s + \left(\frac{x - 1}{2}\right) b\right\},$$

or

$$bx^2 - \{2(S - s) + b\}x = 2tS.$$

Therefore

$$x = \frac{\sqrt{\{2(S - s) + b\}^2 + 8btS} + \{2(S - s) + b\}}{2b},$$

which agrees exactly with the solution as stated in the Bakhshâlî treatise.

"The daily travel [*S*] diminished by the march of the first day [*s*] is doubled; this is increased by the common increment [*b*]. That (sum) multiplied by itself is designated {as the *kṣepa* quantity}. The product of the daily

[1] Datta, "Geometry in the Jaina Cosmography," *Quellen und Studien Zur Ges. d. Math.*, Ab. B, Bd. 1 (1931), pp. 245-254.

travel and the start [t] being multiplied by eight times the common increment, the *kṣepa* quantity is added. The square-root of this {is increased by the *kṣepa* quantity; the sum divided by twice the common increment will give the required number of days}."[1]

Nearly the whole of the detailed working of the particular example in which $S = 5$, $t = 6$, $s = 3$ and $b = 4$, is preserved.[2] It is substantially as follows:

$St = 30$; $S - s = 5 - 3 = 2$; $2(S - s) + b = 8$; $\{2(S - s) + b\}^2 = 64$; $8St = 240$; $8Stb = 960$; $\{2(S - s) + b\}^2 + 8Stb = 1024$; $\sqrt{1024} = 32$; $32 + 8 = 40$; $40 \div 8 = 5 = x$.

For another problem[3] $S = 7$, $t = 5$, $s = 5$, $b = 3$;

then $$x = \frac{7 + \sqrt{889}}{6}.$$

The formula for determining the number of terms (n) of an A.P. whose first term (a), common difference (b) and sum (s) are known, is stated in the form

$$n = \frac{\sqrt{8bs + (2a - b)^2} - (2a - b)}{2b}.$$

The working of the particular example in which $s = 60$, $a = 1$, $b = 1$ is preserved substantially as follows:[4]

$8bs = 480$; $2a = 2$; $2a - b = 1$; $(2a - b)^2 = 1$;

$8bs + (2a - b)^2 = 481$; $n = \frac{1}{2}(-1 + \sqrt{481})$, etc.

Āryabhaṭa I. To find the number of terms of an A.P., Āryabhaṭa I (499) gives the following rule:

[1] *BMs*, Folio 5, recto.

[2] *BMs*, Folio 5, verso; Compare also Kaye's Introduction, pp. 37, 45.

[3] *BMs*, Folio 6, recto and verso.

[4] *BMs*, Folio 65 verso. Working of this example has been continued on folios 56, verso and recto, and 64, recto.

"The sum of the series multiplied by eight times the common difference is added by the square of the difference between twice the first term and the common difference; the square-root (of the result) is diminished by twice the first term and (then) divided by the common difference: half of this quotient plus unity is the number of terms."[1]

That is to say,

$$n = \frac{1}{2}\left\{\frac{\sqrt{8bs + (2a - b)^2} - 2a}{b} + 1\right\}.$$

The solution of a certain interest problem involves the solution of the quadratic

$$tx^2 + px - Ap = 0.$$

Āryabhaṭa gives the value of x in the form[2]

$$x = \frac{\sqrt{Apt + (p/2)^2} - p/2}{t}.$$

Though Āryabhaṭa I has nowhere indicated any method of solving the quadratic, it appears from the above two forms that he followed two different methods in order to make the unknown side of the equation $ax^2 + bx = c$, a perfect square. In one case he multiplied both the sides of the equation by $4a$ and in the other simply by a.

Brahmagupta's Rules. Brahmagupta (628) has given two specific rules for the solution of the quadratic. His first rule is as follows:

"The quadratic: the absolute quantities multiplied by four times the coefficient of the square of the unknown are increased by the square of the coefficient of the middle (*i.e.*, unknown); the square-root of the result being diminished by the coefficient of the middle

[1] *Ā*, ii. 20.
[2] *Ā*, ii. 25; vide Part I, pp. 219f.

and divided by twice the coefficient of the square of the unknown, is (the value of) the middle."[1]

$$i.e., \qquad x = \frac{\sqrt{4ac + b^2} - b}{2a}.$$

The second rule runs as:

"The absolute term multiplied by the coefficient of the square of the unknown is increased by the square of half the coefficient of the unknown; the square-root of the result diminished by half the coefficient of the unknown and divided by the coefficient of the square of the unknown is the unknown."[2]

$$i.e., \qquad x = \frac{\sqrt{ac + (b/2)^2} - (b/2)}{}.$$

The above two methods of Brahmagupta are identical with those employed before him by Āryabhaṭa I (499). The root of the quadratic equation for the number of terms of an A.P. is given by Brahmagupta in the first form:[3]

$$n = \frac{\sqrt{8bs + (2a - b)^2} - (2a - b)}{2b}.$$

For the solution of the quadratic Brahmagupta uses also a third formula which is similar to the one now commonly used. Though it has not been expressly described in any rule, we find its application in a few

[1] *BrSpSi*, xviii. 44. It will be noted that in this rule Brahmagupta has employed the term *madhya* (middle) to imply the simple unknown as well as its coefficient. The original of the term is doubtless connected with the mode of writing the quadratic equation in the form

$$ax^2 + bx + 0 = 0x^2 + 0x + c,$$

so that there are three terms on each side of the equation.

[2] *BrSpSi*, xviii. 45. [3] *BrSpSi*, xii. 18.

instances. One of them is an interest problem: A certain sum (p) is lent out for a period (t_1); the interest accrued (x) is lent out again at this rate of interest for another period (t_2) and the total amount is A. Find x.

The equation for determining x is

$$\frac{t_2}{pt_1}x^2 + x = A.$$

Hence, we have

$$x = \sqrt{\left(\frac{pt_1}{2t_2}\right)^2 + \frac{Apt_1}{t_2}} - \frac{pt_1}{2t_2};$$

which is exactly the form in which Brahmagupta states the result.[1]

There is a certain astronomical problem which involves the quadratic equation[2]

$$(72 + a^2)x^2 \mp 24apx = 144\left(\frac{R^2}{2} - p^2\right),$$

where a = *agrâ* (the sine of the amplitude of the sun), b = *palabhâ* (the equinoctial shadow of a gnomon 12 *aṅguli* long), R = radius, and x = *koṇaśaṅku* (the sine of the altitude of the sun when his altitude is 45°). Dividing out by $(72 + a^2)$, we have

$$x^2 \mp 2mx = n,$$

where

$$m = \frac{12ap}{72 + a^2}, \quad n = \frac{144(R^2/2 - p^2)}{72 + a^2}.$$

Therefore, we have

$$x = \sqrt{m^2 + n} \pm m,$$

as stated by Brahmagupta. This result is given also in the *Sûryasiddhânta*[3] (*c.* 300) and by Śrîpati (1039).[4]

[1] *BrSpSi*, xii. 15. Vide Part I, p. 220.
[2] *BrSpSi*, iii. 54-55.
[3] *SûSi*, iii. 30-1.
[4] *SiŚe*, iv. 74.

Śrîdhara's Rule. Śrîdhara (*c.* 750) expressly indicates his method of solving the quadratic equation. His treatise on algebra is now lost. But the relevant portion of it is preserved in quotations by Bhâskara II[1] and others.[2] Śrîdhara's method is :

"Multiply both the sides (of an equation) by a known quantity equal to four times the coefficient of the square of the unknown; add to both sides a known quantity equal to the square of the (original) coefficient of the unknown: then (extract) the root."[3]

That is, to solve $ax^2 + bx = c,$

we have $4a^2x^2 + 4abx = 4ac,$

or $(2ax + b)^2 = 4ac + b^2.$

Therefore $2ax + b = \sqrt{4ac + b^2}.$

Hence $$x = \frac{\sqrt{4ac + b^2} - b}{2a}.$$

An application of this rule is found in Śrîdhara's *Triśatikâ*, in connection with finding the number of terms of an A.P.[4]

i.e., $$n = \frac{\sqrt{8bs + (2a - b)^2} - 2a + b}{2b},$$

[1] *BBi*, p. 61.

[2] Jñânarâja (1503) in his *Bîjagaṇita* and Sûryadâsa (1541) in his commentary on Bhâskara's *Bîjagaṇita*.

[3] "Caturâhatavargasamai rûpaiḥ pakṣadvayaṁ guṇayet,
Avyaktavargarûpairyuktau pakṣau tato mûlam."

This is the reading of Srîdhara's rule as stated by Jñânarâja and Sûryadâsa and accepted also by Sudhakara Dvivedi. But according to the reading of Kṛṣṇa (*c.* 1580) and Râmakṛṣṇa (*c.* 1648), which has been accepted by Colebrooke, the second line of the verse should be

"Pûrvâvyaktasya kṛteḥ samarûpâṇi kṣipet tayoreva"

or "add to them known quantities equal to the square of the original coefficient of the unknown."

[4] *Triś*, R. 41.

where a is the first term, b the common difference and s the sum of n terms.

Mahâvîra. The only work of Mahâvîra (850) which is available now is the *Ganita-sâra-samgraha*. As it is admittedly devoted to arithmetic we cannot expect to find in it a rule for solving the quadratic. But there are in it several problems whose solutions presuppose a knowledge of the roots of the quadratic. One problem is as follows:

"One-fourth of a herd of camels was seen in the forest; twice the square-root of the herd had gone to the mountain-slopes; three times five camels were on the bank of the river. What was the number of those camels?"[1]

If x be the number of camels in the herd, then

$$\tfrac{1}{4}x + 2\sqrt{x} + 15 = x.$$

Or, in general, the equation to be solved is

$$\frac{a}{b}x + c\sqrt{x} + d = x,$$

or

$$\left(1 - \frac{a}{b}\right)x - c\sqrt{x} = d.$$

Mahâvîra gives the following rule for the solution of this equation:

"Half the coefficient of the root (of the unknown) and the absolute term should be divided by unity minus the fraction (*i.e.*, the coefficient of the unknown). The square-root of the sum of the square of the coefficient of the root (of the unknown) and the absolute term (treated as before) is added to the coefficient of the root (of the unknown treated as before). The sum squared is the (unknown) quantity in this *mûla* type of problems."[2]

[1] *GSS*, iv. 34. [2] *GSS*, iv. 33.

i.e., $$x = \left\{ \frac{c/2}{1 - a/b} + \sqrt{\left(\frac{c/2}{1 - a/b}\right)^2 + \frac{d}{1 - a/b}} \right\}^2,$$

which shows that Mahâvîra employed the modern rule for finding the root of a quadratic. His solution for the interest problem treated by Brahmagupta is exactly the same as that of the latter.[1] We shall presently show that he knew that the quadratic has two roots.

Âryabhaṭa II. The formula for the number of terms (n) of an A.P. whose first term (a), common difference (b) and sum (s) are known is given by Âryabhaṭa II (*c.* 950) as follows :[2]

$$n = \frac{\sqrt{2bs + (a - b/2)^2} - a + b/2}{b},$$

which shows that for solving the quadratic he followed the second method of Âryabhaṭa I and Brahmagupta.

Śrîpati's Rules. Śrîpati (1039) indicates two methods of solving the quadratic. There is a lacuna in our manuscript in the rule describing the first method, but it can be easily recognised to be the same as that of Śrîdhara.

"Multiply by four times the coefficient of the square of the unknown and add the square of the coefficient of the unknown; (then extract) the square-root......... divided by twice the coefficient of the square of the unknown, is said to be (the value of) the unknown."

"Or multiplying by the coefficient of the square of the unknown and adding the square of half the coefficient of the unknown, (extract) the square-root. Then (proceeding) as before, it is diminished by half the coefficient of the unknown and divided by the coefficient

[1] *GSS*, vi. 44. [2] *MSi*, xv. 50.

of the square of the unknown. This (quotient) is said to be (the value of) the unknown."[1]

i.e., $$ax^2 + bx = c,$$

or $$a^2x^2 + abx + (b/2)^2 = ac + (b/2)^2.$$

Therefore $$ax + b/2 = \sqrt{ac + (b/2)^2}.$$

Hence $$x = \frac{\sqrt{ac + (b/2)^2} - b/2}{a}.$$

Bhâskara II's Rules. Bhâskara II (1150) says:

"When the square of the unknown, etc., remain, then, multiplying the two sides (of the equation) by some suitable quantities, other suitable quantities should be added to them so that the side containing the unknown becomes capable of yielding a root (*pada-prada*). The equation should then be formed again with the root of this side and the root of the known side. Thus the value of the unknown is obtained from that equation."[2]

This rule has been further elucidated by the author in his gloss as follows:

"When after perfect clearance of the two sides, there remain on one side the square, etc., of the unknown and on the other side the absolute term only, then, both the sides should be multiplied or divided by some suitable optional quantity; some equal quantities should further be added to or subtracted from both the sides so that the unknown side will become capable of yielding a root. The root of that side must be equal to the root of the absolute terms on the other side. For, by simultaneous equal additions, etc., to the two equal sides the equality remains. So forming an equation again with these roots the value of the unknown is found."[3]

[1] *SiŚe*, xiv. 17-8, 19. [2] *BBi*, p. 59.
[3] *BBi*, p. 61.

It may be noted that in his treatise on arithmetic Bhâskara II has always followed the modern method of dividing by the coefficient of the square of the unknown.[1]

Jñânarâja (1503) and Gaṇeśa (1545) describe the same general methods for solving the quadratic as Bhâskara II.

Elimination of the Middle Term. The method of solving the quadratic was known amongst the Hindu algebraists by the technical designation *madhyamâharaṇa* or "The Elimination of the Middle" (from *madhyama* = middle and *âharaṇa* = removal, or destroying, that is, elimination). The origin of the name will be easily found in the principle underlying the method. By it a quadratic equation which, in its general form, contains three terms and so has a middle term, is reduced to a pure quadratic equation or a simple equation involving only two terms and so having no middle term. Thus the middle term of the original quadratic is eliminated by the method generally adopted for its solution. And hence the name. Bhâskara II has observed, "It is also specially designated by the learned teachers as the *madhyamâharaṇa*. For by it, the removal of one of the two[2] terms of the quadratic, the middle one, (takes place)."[3] The name is, however, employed also in an extended sense so as to embrace the methods for solving the cubic and the biquadratic, where also

[1] *L*, pp. 15ff.

[2] Referring to the two terms on the unknown side of the complete quadratic. Or the text *varga-râśâvekasya* may be rendered as "of one out of the unknown quantity and its square." According to the commentators Sûryadâsa (1541) and Kṛṣṇa (1580), it implies "of one between the two square terms," *viz.*, the square of the unknown and the square of the absolute number.

[3] *BBi*, p. 59.

certain terms are eliminated. It occurs as early as the works of Brahmagupta (628).[1]

Two Roots of the Quadratic. The Hindus recognised early that the quadratic has generally two roots. In this connection Bhâskara II has quoted the following rule from an ancient writer of the name of Padmanâbha whose treatise on algebra is not available now:

"If (after extracting roots) the square-root of the absolute side (of the quadratic) be less than the negative absolute term on the other side, then taking it negative as well as positive, two values (of the unknown) are found."[2]

Bhâskara points out with the help of a few specific illustrations that though these double roots of the quadratic are theoretically correct, they sometimes lead to incongruity and hence should not always be accepted. So he modifies the rule as follows:

"If the square-root of the known side (of the quadratic) be less than the negative absolute term occurring in the square-root of the unknown side, then making it negative as well as positive, two values of the unknown should be determined. This is (to be done) occasionally."[3]

Example 1. "The eighth part of a troop of monkeys, squared, was skipping inside the forest, being delightfully attached to it. Twelve were seen on the hill delighting in screaming and rescreaming. How many were they?"[4]

[1] *BrSpSi*, xviii. 2.

[2] "Vyaktapakṣasya cenmûlamanyapakṣarṇarûpataḥ
Alpaṁ dhanarṇagaṁ kṛtvâ dvividhotpadyate mitiḥ"—*BBi*, p. 67.

[3] *BBi*, p. 59; also compare the author's gloss on the same (p. 61).

[4] *BBi*, p. 65.

Solution. "Here the troop of monkeys is x. The square of the eighth part of this together with 12, is equal to the troop. So the two sides are[1]

$$\tfrac{1}{64}x^2 + 0x + 12 = 0x^2 + x + 0.$$

Reducing these to a common denominator and then deleting the denominator, and also making clearance, the two sides become

$$x^2 - 64x + 0 = 0x^2 + 0x - 768.$$

Adding the square of 32 to both sides and (extracting) square-roots, we get

$$x - 32 = \pm (0x + 16).$$

In this instance the absolute term on the known side is smaller than the negative absolute term on the side of the unknown; hence it is taken positive as well as negative; the two values of x are found to be 48, 16."

Example 2. "The fifth part of a troop of monkeys, leaving out three, squared, has entered a cave; one is seen to have climbed on the branch of a tree. Tell how many are they?"[2]

Solution. "In this the value of the troop is x; its fifth part less three is $\frac{1}{5}x - 3$; squared, $\frac{1}{25}x^2 - \frac{6}{5}x + 9$; this added with the visible (number of monkeys), $\frac{1}{25}x^2 - \frac{6}{5}x + 10$, is equal to the troop. Reducing to a common denominator, then deleting the denominator and making clearance, the two sides become

$$x^2 - 55x + 0 = 0x^2 + 0x - 250.$$

Multiplying these by 4, adding the square of 55, and

[1] We have here followed the modern practice of writing the two sides of an equation in a line with the sign of equality interposed, at the same time, preserving the other salient feature of the Hindu method of indicating the absent terms, if any, by putting zeros as their coefficients.

[2] *BBi*, pp. 65ff.

extracting roots, we get

$$2x - 55 = \pm (0x + 45).$$

In this case also, as in the previous, two values are obtained: 50, 5. But, in this case, the second (value) should not be accepted as it is inapplicable. *People have no faith in the known becoming negative.*"

The implied significance of this last observation is this: If the troop consists of only 5 monkeys, its fifth part will be 1 monkey. How can we then leave out 3 monkeys? Again, how can the remainder be said to have entered the cave? It seems to have also a wider significance.

Example 3. "The shadow of a gnomon of twelve fingers being diminished by a third part of the hypotenuse, becomes equal to fourteen fingers. O mathematician, tell it quickly."[1]

Solution. "Here the shadow is (taken to be) x. This being diminished by a third part of the hypotenuse becomes equal to fourteen fingers. Hence conversely, fourteen being subtracted from it, the remainder, a third of the hypotenuse, is $x - 14$. Thrice this, which is the hypotenuse, is $3x - 42$. The square of it, $9x^2 - 252x + 1764$, is equal to the square of the hypotenuse, $x^2 + 144$. On making equi-clearance, the two sides become

$$8x^2 - 252x + 0 = 0x^2 + 0x - 1620.$$

Multiplying both these sides by 2 and adding the square of 63, the roots are

$$4x - 63 = \pm (0x + 27).$$

On forming an equation with these sides again, and (proceeding) as before, the values of x are 45/2, 9.

[1] *BBi*, pp. 66f.

(Thus) the value of the shadow is 45/2 or 9. The second value of the shadow is less than 14, so, on account of impracticability, it should not be accepted. Hence it has been said 'twofold values occasionally.' This will be an exception to what has been stated in the algebra of Padmanâbha, *viz*..."

Known to Mahâvîra. It has been stated before that Mahâvîra (850) knew that the quadratic has two roots. We shall now substantiate it by the following rules and illustrations from his work.

"One-sixteenth of a collection of peacocks multiplied by itself, was on the mango tree ; $\frac{1}{9}$ of the remainder multiplied by itself together with 14 were on the *tamâla* tree. How many were they ?"[1]

If x be the number of peacocks in the collection, the problem leads to the quadratic equation

$$\frac{x}{16} \times \frac{x}{16} + \frac{15x}{16 \times 9} \times \frac{15x}{16 \times 9} + 14 = x.$$

This is a particular case of the type of equations contemplated by the author

$$\frac{a}{b} x^2 - x + c = 0.$$

The following rule has been given for its solution.

"The quotient of its denominator divided by its numerator, less four times the remainder, is multiplied by that denominator (as divided by the numerator). The square-root of this should be *added to* and *subtracted from* that denominator (as divided by the numerator); half that is the total quantity."[2]

Thus $$x = \frac{b/a \pm \sqrt{(b/a - 4c)b/a}}{2}.$$

[1] *GSS*, iv. 59. [2] *GSS*, iv. 57.

Certain other problems[1] lead to equations of the type

$$\left(\frac{a}{b}x \mp d\right)^2 + c = x.$$

The solution is given as

"Half the denominator divided by its numerator is increased or diminished by the quantity to be subtracted or added. The square of this is diminished by the square of the quantity to be subtracted or added and by the remainder. The square-root of the result *added to* or *subtracted from* the square-root (of the square obtained before) and divided by the fractional part, will be the value (of the unknown)."[2]

$$i.e., \ x = \left\{\left(\frac{b}{2a} \pm d\right) \pm \sqrt{\left(\frac{b}{2a} \pm d\right)^2 - d^2 - c}\right\} \div \frac{a}{b}.$$

We need not add further instances to prove that Mahâvîra recognised both the roots of a quadratic equation.[3] There are, however, a few problems in which he has taken into consideration only one of the roots.[4] For instance, take the equation on page 66,

$$\tfrac{1}{4}x + 2\sqrt{x} + 15 = x,$$

or $$\tfrac{3}{4}x - 2\sqrt{x} = 15.$$

Therefore $$\sqrt{x} = \left\{\tfrac{4}{3} \pm \sqrt{(\tfrac{4}{3})^2 + \frac{4.15}{3}}\right\},$$

or $$\sqrt{x} = (\tfrac{4}{3} \pm \tfrac{14}{3}) = 6 \text{ or } -\tfrac{10}{3}.$$

$$\therefore \quad x = 36.$$

The root can not be negative, hence the negative value of the radical is neglected in the rule.

Brahmagupta. The existence of two roots of a

[1] *GSS*, iv. 62-4. [2] *GSS*, iv. 61.
[3] More instances will be found in *GSS*, vi. 29ff.
[4] See *GSS*, iv. 33-52.

quadratic equation appears to have been known also to Brahmagupta (628). In illustration of his rules for the solution of the quadratic, he has stated two problems involving practically the same equation.

(1) "The square-root of the residue of the revolution of the sun less 2 is diminished by 1, multiplied by 10 and added by 2: when will this be equal to the residue of the revolution of the sun less 1, on Wednesday?"[1]

(2) "When will the square of one-fourth the residue of the exceeding months less three, be equal to the residue of the exceeding months?"[2]

Following Pṛthûdakasvâmî let us take in *Example* 1 the residue of the revolution of the sun to be $x^2 + 2$; then by the question

$$10(x - 1) + 2 = x^2 + 1,$$

or

$$x^2 - 10x = -9.$$

In *Example* 2, put $4x$ for the residue of the exceeding months; then

$$(x - 3)^2 = 4x,$$

or

$$x^2 - 10x = -9.$$

Now, by the second rule of Brahmagupta, retaining both the signs of the radical, we get

$$x = 5 \pm \sqrt{25 - 9} = 9 \text{ or } 1.$$

As shown by Pṛthûdakasvâmî, the first value is taken for *Example* 1 and second value for *Example* 2. Thus it appears that Brahmagupta uses sometimes the positive and at other times the negative sign with the radical. Hence it seems that Brahmagupta knew that a quadratic equation has two roots, though from considerations of utility in his problems, he retains only one of them.

[1] *BrSpSi*, xviii. 49. [2] *BrSpSi*, xviii. 50.

11. EQUATIONS OF HIGHER DEGREES

Cubic and Biquadratic. The Hindus did not achieve much in the solution of the cubic and biquadratic equations. Bhâskara II (1150) attempted the application of the method of the *madhyamâharaṇa* (elimination of the middle) to those equations also so as to reduce them by means of advantageous transformations and introduction of auxiliary quantities to simple and quadratic equations respectively. He thus anticipated one of the modern methods of solving the biquadratic. "If, however," observes Bhâskara II, "due to the presence of the cube, biquadrate, etc., the work (of reduction) cannot proceed any further, after the performance of such operations, for want of a root of the unknown side (of an equation), then the value of the unknown must be obtained by the ingenuity (of the mathematician)."[1] He has given two examples, one of the cubic and the other of the biquadratic, in which such reduction is possible.

Example 1. "What is that number, O learned man, which being multiplied by twelve and increased by the cube of the number, is equal to six times the square of the number added with thirty-five.

Solution. "Here the number is x. This multiplied by twelve and increased by the cube of the number becomes $x^3 + 12x$. It is equal to $6x^2 + 35$. On making clearance, there appears on the first side $x^3 - 6x^2 + 12x$; on the other side 35. Adding negative eight to both the sides and extracting cube-roots, we get

$$x - 2 = 0x + 3.$$

And from this equation the number is found to be 5."[2]

Example 2. "What is that number which being

[1] *BBi*, p. 61. [2] *BBi*, p. 64.

multiplied by 200 and added to the square of the number, and then multiplied by 2 and subtracted from the fourth power of the number will become one myriad less unity? Tell that number if thou be conversant with the operations of analysis.

Solution. "Here the number is x; multiplied by 200 it becomes $200x$; added to the square of the number, becomes $x^2 + 200x$; this being multiplied by two, $2x^2 + 400x$; by this being diminished the fourth power of the number, namely, this x^4, becomes $x^4 - 2x^2 - 400x$. This is equal to a myriad less unity. Equi-clearance having been made, the two sides will be

$$x^4 - 2x^2 - 400x = 0x^4 + 0x^2 + 0x + 9999.$$

Here on adding four hundred x plus unity to the first side, the root can be extracted, but on adding the same to the other side, there will be no root of it. Thus the work (of reduction) does not proceed. Hence here ingenuity (is called for). Here adding to both the sides four times the square of x, four hundred x and unity and then extracting roots, we get

$$x^2 + 0x + 1 = 0x^2 + 2x + 100.$$

Again, forming equation with these and proceeding as before, the value of x is obtained as 11. In similar instances the value of the unknown must be determined by the ingenuity of the mathematician."[1]

Higher Equations. Mahâvîra considered certain simple equations of higher degrees in connection with the treatment of the geometric series. They are of the type

$$(i) \quad ax^n = q,$$

$$(ii) \quad a\frac{x^n - 1}{x - 1} = p;$$

[1] *BBi*, pp. 64f.

where a is the first term of a G. P., q its *guṇadhana*, i.e., $(n+1)$th term, p its sum and x the unknown common ratio.

To solve equation (*i*) Mahâvîra says, "That which on multiplication by itself as many times as the number of terms becomes equal to the *guṇadhana* divided by the first term, is the common ratio."[1]

$$i.e., \qquad x = \sqrt[n]{q/p}.$$

In other words x is the nth root of q/p. But how to find such a root he does not attempt to indicate. His rule for solving an equation of the type (*ii*) is as follows :

"That by which the sum divided by the first term is divisible again and again, subtracting unity every time, is the common ratio."[2]

The method will be better understood from the solution of the following example :

"(Of a certain series in G. P.) the first term is 3, number of terms 6 and sum 4095. What is the common ratio ?"[3]

$$\text{Thus} \qquad 3\frac{x^6 - 1}{x - 1} = 4095,$$

$$\text{or} \qquad 3(x^5 + x^4 + x^3 + x^2 + x + 1) = 4095,$$

a quintic equation. Here dividing 4095 by 3 we get 1365. Now let us try with the divisor 4; we have $(1365 - 1)/4 = 341$; $(341 - 1)/4 = 85$; $(85 - 1)/4 = 21$; $(21 - 1)/4 = 5$; $(5 - 1)/4 = 1$; $(1 - 1)/4 = 0$. So the number 1365 is exhausted on 6 successive divisions by 4, in the way indicated in the rule. Hence $x = 4$. What suggested the method is clearly this :

$$a\frac{x^n - 1}{x - 1} \div a = \frac{x^n - 1}{x - 1}; \quad \frac{x^n - 1}{x - 1} - 1 = x\left(\frac{x^{n-1} - 1}{x - 1}\right);$$

[1] *GSS*, ii. 97. [2] *GSS*, ii. 101.
[3] *GSS*, ii. 102; compare also Rangacarya's note thereto.

which is divisible by x. However, the solution is obtained in every case by trial only.

Mahâvîra has treated some equations of the following general type:

$$a_1\sqrt{b_1x} + a_2\sqrt{b_2(x - a_1\sqrt{b_1x})}$$
$$+ a_3\sqrt{b_3\{(x - a_1\sqrt{b_1x}) - a_2\sqrt{b_2(x - a_1\sqrt{b_1x})}\}}$$
$$+ \dots + R = x;$$

or $$(x - a_1\sqrt{b_1x}) - a_2\sqrt{b_2(x - a_1\sqrt{b_1x})}$$
$$- a_3\sqrt{b_3\{(x - a_1\sqrt{b_1x}) - a_2\sqrt{b_2(x - a_1\sqrt{b_1x})}\}}$$
$$- \dots = R.$$

If there be r terms on the left hand side, then on rationalisation, we shall have an equation of 2^rth degree in x. By proper substitutions, the equation will be ultimately reduced to a quadratic equation of the form

$$X - A\sqrt{BX} = R,$$

whose solution is given by Mahâvîra as

$$X = \left\{\frac{A + \sqrt{A^2 + 4R/B}}{2}\right\}^2 \times B.$$

This result has been termed by him, the "essence" (*sâra*) of the general equation.[1] Mahâvîra gives two problems involving equations of the above type.

(1) "(Of a herd of elephants) nine times the square-root of the two-thirds plus six times the square-root of the three-fifths of the remainder (entered the deep forest); (the remaining) 24 elephants with their round temples wet with the stream of exuding ichor, were seen by me in a forest. How many were the elephants (in

[1] *GSS*, iv. 51, 52.

the herd)?"[1]

If x be the number of elephants in the herd, then by the statement of the problem

$$9\sqrt{\frac{2x}{3}}+6\sqrt{\tfrac{3}{5}\left(x-9\sqrt{\frac{2x}{3}}\right)}+24=x.$$

Put $y=x-9\sqrt{\frac{2x}{3}}$; then the equation becomes

$$y-6\sqrt{3y/5}=24.$$

Therefore $\qquad y=60 \text{ or } \tfrac{48}{5}.$

Hence $\qquad x-9\sqrt{\frac{2x}{3}}=60,$

whence $\qquad x=150, 24.$

Again $\qquad x-9\sqrt{\frac{2x}{3}}=\tfrac{48}{5},$

whence $\qquad x=\tfrac{3}{5}(61\pm 3\sqrt{385}).$

Of the four values of x obtained above, only the value $x=150$ can satisfy all the conditions of the problem; others are inapplicable. That will explain why Mahâvîra has retained in his solution only the positive sign of the radical.

(2) "Four times the square-root of the half of a collection of boars went into a forest where tigers were at play; twice the square-root of the tenth part of the remainder multiplied by 4 went to a mountain; 9 times the square-root of half the remainder went to the bank of a river; boars numbering seven times eight were seen in the forest. Tell their number."[2]

[1] *GSS*, iv. 54-5. [2] *GSS*, iv. 56.

If x be the total number of boars in the collection,

$$4\sqrt{x/2} + 8\sqrt{\tfrac{1}{10}(x - 4\sqrt{x/2})}$$
$$+ 9\sqrt{\tfrac{1}{2}\{(x - 4\sqrt{x/2}) - 8\sqrt{\tfrac{1}{10}(x - 4\sqrt{x/2})}\}}$$
$$+ 56 = x.$$

Put $y = x - 4\sqrt{x/2}$; then

$$y - 8\sqrt{y/10} - 9\sqrt{(y - 8\sqrt{y/10})/2} = 56.$$

Again put $z = y - 8\sqrt{y/10}$; then

$$z - 9\sqrt{z/2} = 56.$$

Therefore $z = \left(\dfrac{9 + \sqrt{81 + 4.2.56}}{2}\right)^2 \times \tfrac{1}{2} = 128.$

Then $y - 8\sqrt{y/10} = 128;$

whence $y = \left(\dfrac{8 + \sqrt{64 + 10.4.128}}{2}\right)^2 \times \tfrac{1}{10} \doteq 160.$

Again $x - 4\sqrt{x/2} = 160;$

hence $x = \left(\dfrac{4 + \sqrt{16 + 4.2.160}}{2}\right)^2 \times \tfrac{1}{2} = 200.$

Note that according to the problem the positive value of the radical has always to be taken.

12. SIMULTANEOUS QUADRATIC EQUATIONS

Common Forms. Various problems involving simultaneous quadratic equations of the following forms have been treated by Hindu writers :

$$\left.\begin{aligned} x - y &= d \\ xy &= b \end{aligned}\right\}\ldots(i) \qquad \left.\begin{aligned} x + y &= a \\ xy &= b \end{aligned}\right\}\ldots(ii)$$

$$\left.\begin{aligned} x^2 + y^2 &= c \\ xy &= b \end{aligned}\right\}\ldots(iii) \qquad \left.\begin{aligned} x^2 + y^2 &= c \\ x + y &= a \end{aligned}\right\}\ldots(iv)$$

For the solution of (*i*) Âryabhaṭa I (499) states the following rule :

"The square-root of four times the product (of two quantities) added with the square of their difference, being added and diminished by their difference and halved gives the two multiplicands."[1]

i.e., $x = \frac{1}{2}(\sqrt{d^2 + 4b} + d), y = \frac{1}{2}(\sqrt{d^2 + 4b} - d).$

Brahmagupta (628) says :

"The square-root of the sum of the square of the difference of the residues and two squared times the product of the residues, being added and subtracted by the difference of the residues, and halved (gives) the desired residues severally."[2]

Nârâyaṇa (1357) writes :

"The square-root of the square of the difference of two quantities plus four times their product is their sum."[3]

"The square of the difference of the quantities together with twice their product is equal to the sum of their squares. The square-root of this result plus twice the product is the sum."[4]

For the solution of (*ii*) the following rule is given by Mahâvîra (850) :

"Subtract four times the area (of a rectangle) from the square of the semi-perimeter; then by *saṅkramaṇa*[5] between the square-root of that (remainder) and the semi-perimeter, the base and the upright are obtained."[6]

[1] *A*, ii. 24. [2] *BrSpSi*, xviii. 99.
[3] *GK*, i. 35. [4] *GK*, i. 36.
[5] Given *a* and *b*, the process of *saṅkramaṇa* is the finding of half their sum and difference, *i.e.*, $\frac{a+b}{2}$ and $\frac{a-b}{2}$ (see pp. 43f).
[6] *GSS*, vii. 129½.

i.e., $x = \frac{1}{2}(a + \sqrt{a^2 - 4b}), y = \frac{1}{2}(a - \sqrt{a^2 - 4b})$.

Nârâyaṇa says :

"The square-root of the square of the sum minus four times the product is the difference."[1]

For (*iii*) Mahâvîra gives the rule :

"Add to and subtract twice the area (of a rectangle) from the square of the diagonal and extract the square-roots. By *saṅkramaṇa* between the greater and lesser of these (roots), the side and upright (are found)."[2]

i.e., $$x = \frac{1}{2}(\sqrt{c + 2b} + \sqrt{c - 2b}),$$
$$y = \frac{1}{2}(\sqrt{c + 2b} - \sqrt{c - 2b}).$$

For equations (*iv*) Âryabhaṭa I writes :

"From the square of the sum (of two quantities) subtract the sum of their squares. Half of the remainder is their product."[3]

The remaining operations will be similar to those for the equations (*ii*); so that

$$x = \frac{1}{2}(a + \sqrt{2c - a^2}), \; y = \frac{1}{2}(a - \sqrt{2c - a^2}).$$

Brahmagupta says :

"Subtract the square of the sum from twice the sum of the squares ; the square-root of the remainder being added to and subtracted from the sum and halved, (gives) the desired residues."[4]

Mahâvîra,[5] Bhâskara II[6] and Nârâyaṇa[7] have also treated these equations.

Nârâyaṇa has given two other forms of simul-

[1] *GK*, i. 35.
[2] *GSS*, vii. 127½.
[3] *Â*, ii. 23.
[4] *BrSpSi*, xviii. 98.
[5] *GSS*, vii. 125½.
[6] *L*, p. 39.
[7] *GK*, i. 37.

taneous quadratic equations, namely,

$$\left.\begin{aligned} x^2+y^2&=c \\ x-y&=d \end{aligned}\right\}\ldots(v) \qquad \left.\begin{aligned} x^2-y^2&=m \\ xy&=b \end{aligned}\right\}\ldots(vi)$$

For the solution of (v) he gives the rule :

"The square-root of twice the sum of the squares decreased by the square of the difference is equal to the sum."[1]

i.e., $$x+y=\sqrt{2c-d^2}.$$

Therefore

$$x=\tfrac{1}{2}(\sqrt{2c-d^2}+d),\quad y=\tfrac{1}{2}(\sqrt{2c-d^2}-d).$$

For (vi) Nârâyaṇa writes :

"Suppose the square of the product as the product (of two quantities) and the difference of the squares as their difference. From them by *saṅkrama* will be obtained the (square) quantities. Their square-roots severally will give the quantities (required)."[2]

We have

$$\left.\begin{aligned} x^2-y^2&=m \\ x^2y^2&=b^2 \end{aligned}\right\}$$

These are of the form (i). Therefore

$$x^2=\tfrac{1}{2}(\sqrt{m^2+4b^2}+m),\quad y^2=\tfrac{1}{2}(\sqrt{m^2+4b^2}-m).$$

Whence we get the values of x and y.

Rule of Dissimilar Operations. The process of solving the following two particular cases of simultaneous quadratic equations was distinguished by most Hindu mathematicians by the special designation ***viṣama-karma***[3] (dissimilar operation) :

[1] *GK*, i. 33. [2] *GK*, i. 34.

[3] The name *viṣama-karma* originated obviously in contradistinction to the name *saṅkramaṇa*. This is evident from the term *viṣama-saṅkramaṇa* used by Mahâvîra for *viṣama-karma*.

$$\left.\begin{array}{r} x^2 - y^2 = m \\ x - y = n \end{array}\right\} \dots (i) \qquad \left.\begin{array}{r} x^2 - y^2 = m \\ x + y = p \end{array}\right\} \dots (ii)$$

These equations are found to have been regarded by them as of fundamental importance. The solutions given are:

for (i) $\quad x = \frac{1}{2}\left(\frac{m}{n} + n\right), \quad y = \frac{1}{2}\left(\frac{m}{n} - n\right);$

for (ii) $\quad x = \frac{1}{2}\left(p + \frac{m}{p}\right), \quad y = \frac{1}{2}\left(p - \frac{m}{p}\right).$

Thus Brahmagupta says:

"The difference of the squares (of the unknowns) is divided by the difference (of the unknowns) and the quotient is increased and diminished by the difference and divided by two; (the results will be the two unknown quantities); (this is) dissimilar operation."[1]

The same rule is restated by him on a different occasion in the course of solving a problem.

"If then the difference of their squares, also the difference of them (are given): the difference of the squares is divided by the difference of them, and this (latter) is added to and subtracted from the quotient and then divided by two; (the results are) the residues; whence the number of elapsed days (can be found)."[2]

Mahâvîra states:

"The *saṅkramaṇa* of the divisor and the quotient of the two quantities is dissimilar (operation); so it is called by those who have reached the end of the ocean of mathematics."[3]

Similar rules are given also by other writers.[4]

[1] *BrSpSi*, xviii. 36. [2] *BrSpSi*, xviii. 97.
[3] *GSS*, vi. 2.
[4] *MSi*, xv. 22; *SiŚe*, xiv. 13; *L*, pp. 13, 37; *GK*, i. 32.

Mahâvîra's Rules. Mahâvîra (850) has treated certain problems involving the simultaneous quadratic equations :

$$u + x = a, \qquad urw = \alpha x,$$
$$u + y = b. \qquad usw = \alpha y.$$

Here $$\frac{r}{s} = \frac{x}{y} = \frac{a - u}{b - u}.$$

Therefore $$u = \frac{rb - sa}{r - s}.$$

Hence $x = \left(\frac{a - b}{r - s}\right)r$, $y = \left(\frac{a - b}{r - s}\right)s$, $w = \left(\frac{a - b}{rb - sa}\right)\alpha$.

In the above equations x, y are the interests accrued on the principal u in the periods r, s respectively and w is the rate of interest per α.

Mahâvîra states the result thus :

"The difference of the mixed sums [a, b] multiplied by each other's periods [r, s], being divided by the difference of the periods, the quotient is known as the principal [u]."[1]

Again, there are problems involving the equations:

$$u + x = p, \qquad uxw = \alpha m,$$
$$u + y = q. \qquad uyw = \alpha n.$$

Where x, y are the periods for which the principal u is lent out at the rate of interest w per α and m, n are the respective interests.

Here $$\frac{m}{n} = \frac{x}{y} = \frac{p - u}{q - u}.$$

Therefore $$u = \frac{mq - np}{m - n}.$$

[1] *GSS*, vi. 47.

Hence $x = \left(\frac{p-q}{m-n}\right) m, \quad y = \left(\frac{p-q}{m-n}\right) n,$

$$w = \frac{\alpha(m-n)^2}{(p-q)(mq-np)}.$$

Mahâvîra gives the rule :

"On the difference of the mixed sums multiplied by each other's interests, being divided by the difference of the interests, the quotient, the wise men say, is the principal."[1]

13. INDETERMINATE EQUATIONS OF THE FIRST DEGREE

General Survey. The earliest Hindu algebraist to give a treatment of the indeterminate equation of the first degree is Āryabhaṭa I (born 476). He gave a method for finding the general solution in positive integers of the simple indeterminate equation

$$by - ax = c$$

for integral values of a, b, c and further indicated how to extend it to get positive integral solutions of simultaneous indeterminate equations of the first degree. His disciple, Bhâskara I (522), showed that the same method might be applied to solve $by - ax = -c$ and further that the solution of this equation would follow from that of $by - ax = -1$. Brahmagupta and others simply adopted the methods of Āryabhaṭa I and Bhâskara I. About the middle of the tenth century of the Christian Era, Āryabhaṭa II improved them by pointing out how the operations can in certain cases be abridged considerably. He also noticed the cases of failure of the methods for an equation of the form

[1] *GSS*, vi. 51.

$by - ax = \pm c$. These results reappear in the works of later writers.[1]

Its Importance. It has been observed before that the subject of indeterminate analysis of the first degree was considered so important by the ancient Hindu algebraists that the whole science of algebra was once named after it. That high estimation of the subject continued undiminished amongst the later Hindu mathematicians. Âryabhaṭa II enumerates it distinctively along with the sciences of arithmetic, algebra, and astronomy.[2] So did Bhâskara II and others. As has been remarked by Gaṇeśa,[3] the separate mention of the subject of indeterminate analysis of the first degree is designed to emphasize its difficulty and importance. On account of its special importance, the treatment of this subject has been included by Bhâskara II in his treatise of arithmetic also, though it belongs particularly to algebra.[4] It is also noteworthy that there is a work exclusively devoted to the treatment of this subject. Such a special treatise is a very rare thing in the mathematical literature of the ancient Hindus. This work, entitled *Kuṭṭâkâra-śiromaṇi*,[5] is by one Devarâja, a commentator of Âryabhaṭa I.

[1] For "India's Contribution to the Theory of Indeterminate Equations of the First Degree," see the comprehensive article of Professor Sarada Kanta Ganguly in *Journ. Ind. Math. Soc.*, XIX, 1931, *Notes and Questions*, pp. 110-120, 129-142; see also XX, 1932, *Notes and Questions*. Compare also the Dissertation of D. M. Mehta on "Theory of simple continued fractions (with special reference to the history of Indian Mathematics)."

[2] *MSi*, i. 1.

[3] Vide his commentary on the *Lîlâvatî* of Bhâskara II.

[4] Bhâskara's treatment of the pulveriser in his *Bîjagaṇita* is repeated nearly word for word in his *Lîlâvatî*.

[5] There are four manuscript copies of this work in the Oriental Library, Mysore.

Three Varieties of Problems. Problems whose solutions led the ancient Hindus to the investigation of the simple indeterminate equation of the first degree were distinguished broadly into three varieties. The problem of one variety is to find a number (N) which being divided by two given numbers (a, b) will leave two given remainders (R_1, R_2). Thus we have

$$N = ax + R_1 = by + R_2.$$

Hence $$by - ax = R_1 - R_2.$$

Putting $$c = R_1 \sim R_2,$$

we get $$by - ax = \pm c$$

the upper or lower sign being taken according as $R_1 >$ or $< R_2$. In a problem of the second kind we are required to find a number (x) such that its product with a given number (α) being increased or decreased by another given number (γ) and then divided by a third given number (β) will leave no remainder. In other words we shall have to solve

$$\frac{\alpha x \pm \gamma}{\beta} = y$$

in positive integers. The third variety of problems similarly leads to equations of the form

$$by + ax = \pm c.$$

Terminology. The subject of indeterminate analysis of the first degree is generally called by the Hindus *kuṭṭaka*, *kuṭṭâkâra*, *kuṭṭîkâra* or simply *kuṭṭa*. The names *kuṭṭâkâra* and *kuṭṭa* occur as early as the *Mahâ-Bhâskarîya* of Bhâskara I (522).[1] In the commentary of the *Âryabhaṭîya* by this writer we find the terms *kuṭṭaka* and *kuṭṭâkâra*. Brahmagupta has used *kuṭṭaka*,[2] *kuṭṭâkâra*,[3] and *kuṭṭa*.[4] Mahâvîra, it appears, had a

[1] *MBh*, i. 41, 49.

[2] *BrSpSi*, xviii. 2, 11, etc.

[3] *BrSpSi*, xviii. 6, 15, etc.

[4] *BrSpSi*, xviii. 20, 25, etc.

preferential liking for the name *kuṭṭîkâra*.[1]

In a problem of the first variety the quantities (a, b) are called "divisors" (*bhâgahâra*, *bhâjak*, *cheda*, etc.) and (R_1, R_2) "remainders" (*agra*, *śeṣa*, etc.), while in a problem of the second variety, β is ordinarily called the "divisor" and γ the "interpolator" (*kṣepa*, *kṣepaka*, etc.); here α is called the "dividend" (*bhâjya*), the unknown quantity to be found (x) the "multiplier" (*guṇaka*, *guṇakâra*, etc.) and y the quotient (*phala*). The unknown (x) has been sometimes called by Mahâvîra as *râśi* (number) implying "an unknown number."[2]

Origin of the name. The Sanskrit words *kuṭṭa*, *kuṭṭaka*, *kuṭṭâkâra* and *kuṭṭîkâra* are all derived from the root *kuṭṭ* "to crush", "to grind," "to pulverise" and hence etymologically they mean the act or process of "breaking", "grinding", "pulverising" as well as an instrument for that, that is, "grinder", "pulveriser". Why the subject of the indeterminate analysis of the first degree came to be designated by the term *kuṭṭaka* is a question which will be naturally asked. Gaṇeṣa (1545) says: "*Kuṭṭaka* is a term for the multiplier, for multiplication is admittedly called by words importing 'injuring,' 'killing.' A certain given number being multiplied by another (unknown quantity), added or subtracted by a given interpolator and then divided by a given divisor leaves no remainder; that multiplier is the *kuṭṭaka*: so it has been said by the ancients. This is a special technical term."[3] The same explanation as to the origin of the name *kuṭṭaka* has been offered by Sûryadâsa (1538), Kṛṣṇa (*c.* 1580) and Raṅganâtha (1602).[4]

[1] *GSS*, vi. 79½, etc. [2] *GSS*, vi. 115½ff.

[3] Vide his commentary on the *Lîlâvatî* of Bhâskara II.

[4] Vide the commentaries of Sûryadâsa on *Lîlâvatî* and *Bîjagaṇita*, of Kṛṣṇa on *Bîjagaṇita*, and of Raṅganâtha on *Siddhânta-śiromaṇi*.

But it is one-sided inasmuch as it has admittedly in view a problem of the second variety where we have indeed to find an unknown multiplier. But the rules of the earlier algebraists such as Âryabhaṭa I and Brahmagupta were formulated with a view to the solution of a problem of the first variety. So the considerations which led those early writers to adopt the name *kuṭṭaka* must have been different. Mahâvîra has once stated that, according to the learned, *kuṭṭîkâra* is another name for "the operation of *prakṣepaka*" (lit., throwing, scattering, implying division into parts).[1] In fact, his writing led his translator to interpret *kuṭṭîkâra* as "proportionate division", "a special kind of division or distribution."[2] Bhâskara I, who had in view a problem of the second variety, once remarked, "the number is obtained by the operation of pulverising (*kuṭṭana*), when it is desired to get the multiplier (*guṇakâra*). . . . "[3] It will be presently shown that the Hindu method of solving the equation $by - ax = \pm c$ is essentially based on a process of deriving from it successively other similar equations in which the values of the coefficients (a, b) become smaller and smaller.[4] Thus the process is indeed the same as that of breaking a whole thing into smaller pieces. In our opinion, it is this that led the ancient mathematicians to adopt the name *kuṭṭaka* for the operation.

Preliminary Operations. It has been remarked by most of the writers that in order that an equation

[1] "Prakṣepaka-karaṇamidaṁ.....kuttîkâro budhaissamuddiṣṭaṃ"—*GSS*, vi. 79½.

[2] Vide *GSS* (English translation), pp. 117, 300.

[3] "Kṛta-kuṭṭana-labdha-râśimeṣâṃ
Guṇakâraṃ samuśanti......"—*MBh*, i. 48.

[4] It has been expressly stated by Sûryadeva Yajvâ that the process must be continued "yâvaddharabhâjyayoralpatâ."

of the form

$$by - ax = \pm c \quad \text{or} \quad by + ax = \pm c$$

may be solvable, the two numbers a and b must not have a common divisor; for, otherwise, the equation would be absurd, unless the number c had the same common divisor. So before the rules adumbrated hereafter can be applied, the numbers a, b, c must be made prime (*dṛḍha* = firm, *niccheda* = having no divisor, *nirapavarta* = irreducible) to each other.

Thus Bhâskara I observes:

"The dividend and divisor will become prime to each other on being divided by the residue of their mutual division. The operation of the pulveriser should be considered in relation to them."[1]

Brahmagupta says:

"Divide the multiplier and the divisor mutually and find the last residue; those quantities being divided by the residue will be prime to each other."[2]

Āryabhaṭa II has made the preliminary operations in successive stages. These will be described later on.[3]

Śrîpati states:

"The dividend, divisor and interpolator should be divided by their common divisor, if any, so that it may be possible to apply the method to be described."[4]

"If the dividend and divisor have a common divisor, which is not a divisor of the interpolator then the problem would be absurd."[5]

Bhâskara II writes:

"As preparatory to the method of the pulveriser,

[1] *MBh*, i. 41. [2] *BrSpSi*, xviii. 9.
[3] *Vide infra*, p. 104. [4] *SiŚe*, xiv. 22.
[5] *SiŚe*, xiv. 26.

the dividend, divisor and interpolator must be divided by a common divisor, if possible. If the number by which the dividend and divisor are divisible, does not divide the interpolator then the problem is absurd. The last residue of the mutual division of two numbers is their common divisor. The dividend and divisor, being divided by their common divisor, become prime to each other."[1]

Rules similar to these have been given also by Nârâyaṇa,[2] Jñânarâja and Kamalâkara.[3] So in our subsequent treatment of the Hindu methods for the solution in positive integers of the equation $by \pm ax = \pm c$, we shall always take, unless otherwise stated, a, b prime to each other.

Solution of $by - ax = \pm c$

Âryabhaṭa I's Rule. The rule of Âryabhaṭa I (499)[4] is rather obscure inasmuch as all the operations intended to be carried out have not been described fully and clearly. So it has been misunderstood by many writers.[5] Following the interpretation of the rule by Bhâskara I (525), a direct disciple of Âryabhaṭa I, Bibhutibhusan Datta has recently given the following translation:[6]

[1] *L*, p. 76; *BBi*, pp. 24f.

[2] *NBi*, I, R. 53-4.

[3] *SiTVi*, xiii. 179ff.

[4] *Â*, ii. 32-3.

[5] L. Rodet, "Leçons de calcul d'Âryabhatta," *JA*, XIII, 1878, pp. 303ff; G. R. Kaye, "Notes on Indian Mathematics. No. 2—Âryabhaṭa," *JASB*, IV, 1908, pp. 111ff; *BCMS*, IV, p. 55; N. K. Mazumdar, "Aryyabhatta's rule in relation to Indeterminate Equations of the First Degree," *BCMS*, III, pp 11-9; P. C. Sen Gupta, "Âryabhaṭîyam," *Jour. Dept. Let. Cal. Univ.*, XVI, 1927; reprint, p. 27.; S. K. Ganguly, *BCMS*, XIX, 1928, pp. 170ff; W. E. Clark, *Âryabhaṭîya of Âryabhaṭa*, Chicago, 1930, pp. 42ff.

[6] Bibhutibhusan Datta, "Elder Âryabhaṭa's rule for the solution of indeterminate equations of the first degree," *BCMS*, XXIV, 1932, pp. 35-53.

"Divide the divisor corresponding to the greater remainder by the divisor corresponding to the smaller remainder. The residue (and the divisor corresponding to the smaller remainder) being mutually divided, the last residue should be multiplied by such an optional integer that the product being added (in case the number of quotients of the mutual division is even) or subtracted (in case the number of quotients is odd) by the difference of the remainders (will be exactly divisible by the last but one remainder. Place the quotients of the mutual division successively one below the other in a column; below them the optional multiplier and underneath it the quotient just obtained). Any number below (*i.e.*, the penultimate) is multiplied by the one just above it and then added by that just below it. Divide the last number (obtained by doing so repeatedly[1]) by the divisor corresponding to the smaller remainder; then multiply the residue by the divisor corresponding to the greater remainder and add the greater remainder. (The result will be) the number corresponding to the two divisors."

He has further shown that it can be rendered also as follows:

"Divide the divisor corresponding to the greater remainder by the divisor corresponding to the smaller remainder. The residue (and the divisor corresponding to the smaller remainder) being mutually divided (until the remainder becomes zero), the last quotient should be multiplied by an optional integer and then added (in case the number of quotients of the mutual division is even) or subtracted (in case the number of quotients is odd) by the difference of the remainders. (Place the other quotients of the mutual division succes-

[1] The process implied here is shown in detail in the working of the example on pages 113f.

sively one below the other in a column; below them the result just obtained and underneath it the optional integer). Any number below (*i.e.*, the penultimate) is multiplied by the one just above it and then added by that just below it. Divide the last number (obtained by doing so repeatedly) by the divisor corresponding to the smaller remainder; then multiply the residue by the divisor corresponding to the greater remainder and add the greater remainder. (The result will be) the number corresponding to the two divisors."

Āryabhaṭa's problem is : To find a number (N) which being divided by two given numbers (a, b) will leave two given remainders (R_1, R_2).[1] This gives:

$$N = ax + R_1 = by + R_2.$$

Denoting as before by c the difference between R_1 and R_2, we get

$$(i) \quad by = ax + c, \text{ if } R_1 > R_2,$$

or

$$(ii) \quad ax = by + c, \text{ if } R_2 > R_1$$

the equation being so written as to keep c always positive. Hence the problem now reduces to making either

$$\frac{ax + c}{b} \quad \text{or} \quad \frac{by + c}{a}$$

according as $R_1 > R_2$ or $R_2 > R_1$, a positive integer. So Āryabhaṭa says: "Divide the divisor corresponding to the greater remainder etc."

[1] It has already been stated (p. 90) that in a problem of the first variety which gives an equation of the above form (and in which $R_1 > R_2$).

a = divisor corresponding to greater remainder,
b = divisor corresponding to lesser remainder,
R_1 = greater remainder,
R_2 = lesser remainder.

Suppose $R_1 > R_2$; then the equation to be solved will be

$$ax + c = by \qquad \text{(I)}$$

a, b being prime to each other.

Let

$$
\begin{array}{llllll}
b)\,a\,(q & & & & & \\
\underline{bq} & & & & & \\
r_1)\,b\,(q_1 & & & & & \\
 & \underline{r_1 q_1} & & & & \\
 & r_2)\,r_1\,(q_2 & & & & \\
 & & \underline{r_2 q_2} & & & \\
 & & r_3 & & & \\
 & & \underline{\ldots\ldots} & & & \\
 & & r_{m-1})\,r_{m-2}\,(q_{m-1} & & & \\
 & & & \underline{r_{m-1}q_{m-1}} & & \\
 & & & r_m)\,r_{m-1}\,(q_m & & \\
 & & & & \underline{r_m q_m} & \\
 & & & & r_{m+1} &
\end{array}
$$

Then, we get[1]

$$
\begin{aligned}
a &= bq + r_1, \\
b &= r_1 q_1 + r_2, \\
r_1 &= r_2 q_2 + r_3, \\
r_2 &= r_3 q_3 + r_4, \\
\ldots & \quad \ldots \\
r_{m-2} &= r_{m-1} q_{m-1} + r_m, \\
r_{m-1} &= r_m q_m + r_{m+1}.
\end{aligned}
$$

Now, substituting the value of a in the given equation (I), we get

$$by = (bq + r_1)x + c.$$

Therefore

$$y = qx + y_1,$$

[1] When $a < b$, we shall have $q = 0$, $r_1 = a$.

where $$by_1 = r_1 x + c.$$

In other words, since $a = bq + r_1$, on putting

$$y = qx + y_1 \tag{1}$$

the given equation (I) reduces to

$$by_1 = r_1 x + c. \tag{I. 1}$$

Again, since $b = r_1 q_1 + r_2$,

putting similarly $x = q_1 y_1 + x_1$.

the equation (I. 1) can be further reduced to

$$r_1 x_1 = r_2 y_1 - c \tag{I. 2}$$

and so on.

Writing down the successive values and reduced equations in columns, we have

(1)	$y = qx + y_1,$	$by_1 = r_1 x + c,$	(I. 1)
(2)	$x = q_1 y + x_1,$	$r_1 x_1 = r_2 y_1 - c,$	(I. 2)
(3)	$y_1 = q_2 x_1 + y_2,$	$r_2 y_2 = r_3 x_1 + c,$	(I. 3)
(4)	$x_1 = q_3 y_2 + x_2,$	$r_3 x_2 = r_4 y_2 - c,$	(I. 4)
(5)	$y_2 = q_4 x_2 + y_3,$	$r_4 y_3 = r_5 x_2 + c,$	(I. 5)
(6)	$x_2 = q_5 y_3 + x_3,$	$r_5 x_3 = r_6 y_3 - c,$	(I. 6)
	....	....	
$(2n-1)$	$y_{n-1} = q_{2n-2} x_{n-1} + y_n,$	$r_{2n-2} y_n = r_{2n-1} x_{n-1} + c,$	(I. $\overline{2n-1}$)
$(2n)$	$x_{n-1} = q_{2n-1} y_n + x_n,$	$r_{2n-1} x_n = r_{2n} y_n - c,$	(I. $2n$)
$(2n+1)$	$y_n = q_{2n} x_n + y_{n+1},$	$r_{2n} y_{n+1} = r_{2n+1} x_n + c,$	(I. $\overline{2n+1}$)

Now the mutual division can be continued either (*i*) to the finish or (*ii*) so as to get a certain number of quotients and then stopped. In either case the number of quotients found, neglecting the first one (q), as is usual with Āryabhaṭa, may be even or odd.

Case i. First suppose that the mutual division is continued until the zero remainder is obtained. Since a, b are prime to each other, the last but one remainder is unity.

Subcase (*i.* 1). Let the number of quotients be *even.* We then have

$$r_{2n} = 1, \; r_{2n+1} = 0, \; q_{2n} = r_{2n-1}.$$

The equations (I. $2n$) and (I. $\overline{2n+1}$), therefore, become

$$y_n = q_{2n}.x_n + c$$

and
$$y_{n+1} = c$$

respectively. Giving an arbitrary integral value (t) to x_n, we get an integral value of y_n. From that we can find the value of x_{n-1} by ($2n$). Proceeding backwards step by step we ultimately find the values of x and y in positive integers. So that the equation (I) is solved.

Subcase (*i.* 2). If the number of quotients be *odd*, we shall have

$$r_{2n-1} = 1, \; r_{2n} = 0, \; q_{2n-1} = r_{2n-2}.$$

The equations ($2n+1$) and (I. $\overline{2n+1}$) will then be absent and the equations (I. $\overline{2n-1}$) and (I. $2n$) will be reduced respectively to

$$x_{n-1} = q_{2n-1}.y_n - c$$

and
$$x_n = -c.$$

Giving an arbitrary integral value (t') to y_n we get an integral value of x_{n-1}. Then proceeding backwards as before we can calculate the values of x and y.

Case ii. Next suppose that the mutual division is stopped after having obtained an even or odd number of quotients.

Subcase (*ii.* 1). If the number of quotients obtained be *even*, the reduced form of the original equation is

$$r_{2n}.y_{n+1} = r_{2n+1}x_n + c,$$

or
$$y_{n+1} = \frac{r_{2n+1}x_n + c}{r_{2n}}.$$

Giving a suitable integral value (t) to x_n as will make

$$y_{n+1} = \frac{r_{2n+1}t + c}{r_{2n}} = \text{an integral number},$$

we get an integral value for y_n by $(2n+1)$. The values of x and y can then be calculated by proceeding as before.

Subcase (*ii.* 2). If the number of quotients be *odd*, the reduced form of the quotient is

$$r_{2n-1}x_n = r_{2n}y_n - c,$$

or
$$x_n = \frac{r_{2n}y_n - c}{r_{2n-1}}.$$

Putting $y_n = t'$, where t' is an integer, such that

$$x_n = \frac{r_{2n}t' - c}{r_{2n-1}} = \text{a whole number},$$

we get an integral value of x_{n-1} by $(2n)$. Whence can be calculated the values of x and y in integers.

If $x = \alpha$, $y = \beta$ be the least integral solution of $ax + c = by$, we shall have

$$a\alpha + c = b\beta.$$

Therefore $\quad a(bm + \alpha) + c = b(am + \beta),$

m being any integer. Therefore, in general,

$$x = bm + \alpha.$$

But we have calculated before that

$$x = q_1y_1 + x_1;$$

$$\therefore \quad q_1y_1 + x_1 = bm + \alpha.$$

Thus it is found that the minimum value α of x is equal to the remainder left on dividing its calculated value by b. Whence we can calculate the minimum value of $N(= a\alpha + R_1)$. This will explain the *rationale* of the operations described in the latter portion of the rule of Âryabhaṭa I.

Bhâskara I's Rules. Bhâskara I (522) writes:

"Set down the dividend above and the divisor below. Write down successively the quotients of their mutual division, one below the other, in the form of a chain. Now find by what number the last remainder should be multiplied, such that the product being subtracted by the (given) residue (of the revolution) will be exactly divisible (by the divisor corresponding to that remainder). Put down that optional number below the chain and then the (new) quotient underneath. Then multiply the optional number by that quantity which stands just above it and add to the product the (new) quotient (below). Proceed afterwards also in the same way. Divide the upper number (*i.e.*, the multiplier) obtained by this process by the divisor and the lower one by the dividend; the remainders will respectively be the desired *ahargaṇa* and the revolutions."[1]

The equation contemplated in this rule is[2]

$$\frac{ax - c}{b} = \text{a positive integer.}$$

This form of the equation seems to have been chosen by Bhâskara I deliberately so as to supplement the form of Āryabhaṭa I in which the interpolator is always made positive by necessary transposition. Further b is taken to be greater than a, as is evident from the following rule. So the first quotient of mutual division of a by b is always zero. This has not been taken into consideration. Also the number of quotients in the chain is taken to be even.

[1] *MBh*, i. 42-4.

The above rule has been formulated with a view to its application in astronomy.

[2] As already stated on p. 90, when the equation is stated in this *second* form

a = dividend,
b = divisor,
c = interpolator,
x = multiplier,
y = quotient.

He further observes:

"When the dividend is greater than the divisor, the operations should be made in the same way (*i.e.*, according to the method of the pulveriser) after deleting the greatest multiple of the divisor (from the dividend). Multiply the (new) multiplier thus obtained by that multiple and add the (new) quotient; the result will be the quotient here (required)."[1]

That is to say, if in the equation

$$ax \pm c = by,$$

$a = mb + a'$, we may neglect the portion mb of the dividend and proceed at once with the solution of

$$a'x \pm c = by.$$

Let $x = \alpha$, $y = \beta$ be a solution of this equation. Then

$$a'\alpha \pm c = b\beta;$$

$$\therefore \quad (mb + a')\alpha \pm c = b(m\alpha + \beta),$$

or

$$a\alpha \pm c = b(m\alpha + \beta).$$

Hence $x = \alpha$, $y = m\alpha + \beta$ is a solution of the given equation.

Brahmagupta's Rules. For the solution of Ārya-bhaṭa's problem Brahmagupta (628) gives the following rule:

"What remains when the divisor corresponding to the greater remainder is divided by the divisor corresponding to the smaller remainder—that (and the latter divisor) are mutually divided and the quotients are severally set down one below the other. The last residue (of the reciprocal division after an even number of quotients has been obtained[2]) is multiplied by

[1] *MBh*, i. 47.

[2] Compare the next rule: "Such is the process when the quotients (of mutual division) are even etc."

such an optional integer that the product being added with the difference of the (given) remainders will be exactly divisible (by the divisor corresponding to that residue). That optional multiplier and then the (new) quotient just obtained should be set down (underneath the listed quotients). Now, proceeding from the lowermost number (in the column), the penultimate is multiplied by the number just above it and then added by the number just below it. The final value thus obtained (by repeating the above process) is divided by the divisor corresponding to the smaller remainder. The residue being multiplied by the divisor corresponding to the greater remainder and added to the greater remainder will be the number in view."[1]

He further observes:

"Such is the process when the quotients (of mutual division) are even in number. But if they be odd, what has been stated before as negative should be made positive or as positive should be made negative."[2]

Regarding the direction for dividing the divisor corresponding to the greater remainder by the divisor corresponding to the smaller remainder, Pṛthûdakasvâmî (860) observes that it is not absolute, rather optional; so that the process may be conducted in the same way by starting with the division of the divisor corresponding to the smaller remainder by the divisor corresponding to the greater remainder. But in this case of inversion of the process, he continues, the difference of the remainders must be made negative.

That is to say, the equation

$$by = ax + c$$

can be solved by transforming it first to the form

$$ax = by - c,$$

[1] *BrSpSi*, xviii. 3-5. [2] *BrSpSi*, xviii. 13.

so that we shall have to start with the division of b by a.

Mahâvîra's Rules. Mahâvîra (850) formulates his rules with a view to the solution of

$$\frac{ax \pm c}{b} = y,$$

in positive integers. He says:

"Divide the coefficient of the unknown by the given divisor (mutually); reject the first quotient and then set down the other quotients of mutual division one below the other. When the residue has become sufficiently small, multiply it by an optional number such that the product, being combined with the interpolator, which if positive must be made negative (and *vice versa*) in case (the number of quotients retained is) odd, will be exactly divisible (by the divisor corresponding to that residue). Place that optional number and the resulting quotient in order under the chain of quotients. Now add the lowermost number to the product of the next two upper numbers. The number (finally obtained by this process) being divided by the given divisor, (the remainder will be the least value of the unknown)."[1]

This method has been redescribed by Mahâvîra in a slightly modified form. Here he continues the mutual division until the remainder zero is obtained and further takes the optional multiplier to be zero.

"With the dividend, divisor and remainder reduced (by their greatest common factor the operations should be performed). Reject the first quotient and set down the other quotients of mutual division (one below the other) and underneath them the zero[2] and the given remainder

[1] *GSS*, vi. 115½ (first portion).

[2] We have emended *sâgra* of the printed text to *khâgra*.

(as reduced) in succession. The remainder, being multiplied by positive or negative as the number of quotients is even or odd, should be added to the product of the next two upper numbers. The number (finally obtained by the repeated application of this process) whether positive or negative, being divided by the divisor, the remainder will be (the least value of) the multiplier."[1]

Āryabhaṭa II. The details of the process adopted by Āryabhaṭa II (950) in finding the general solution of $(ax \pm c)/b = y$ in positive integers have been described by him thus:

"Set down the dividend, interpolator and divisor as stated (in a problem): this is the *first operation.*

"Divide them by their greatest common divisor so as to make them without a common factor: this is the *second operation.*

"Divide the dividend and interpolator by their greatest common divisor: the *third operation.*

"Divide the interpolator and divisor by their greatest common divisor: the *fourth operation.*

"Divide the dividend and interpolator, then the interpolator (thus reduced) and divisor by their respective different greatest common divisors: the *fifth operation.*

"On forming the chain from these (reduced numbers), if the remainder becomes unity, then the object (of solving the problem) will be realised; but if the remainder in it be zero, the questioner does not know the method of the pulveriser.

"Divide the (reduced) dividend and divisor reciprocally until the remainder becomes unity. (The quotients placed one below the other successively will form)

[1] *GSS*, vi. 136½ (first portion). Our interpretation differs from those of Rangacharya and Ganguly.

the (auxiliary) chain. Note down whether the number of quotients is even or odd. Multiply by the ultimate the number just above it and then add unity. The chain formed on replacing the penultimate by this result is the corrected one. Multiply by the un-destroyed (*i.e.*, corrected) penultimate the number just above it, then add the ultimate number; (now) destroy the ultimate. On proceeding thus (repeatedly) we shall finally obtain two numbers which are (technically) called *kuṭṭa*. I shall speak (later on) of those two quantities as obtained in the case of an odd number of quotients. If on dividing the dividend by the divisor once only the residue becomes unity, then the quotient is known to be the upper *kuṭṭa* and the remainder (*i.e.*, unity) the lower *kuṭṭa*.

"The upper and lower *kuṭṭa* thus obtained, being both multiplied by the interpolator of the given equation and then divided respectively by its dividend and divisor, the residues will be the quotient and multiplier respectively.

"In the case of the third operation (having been performed before) multiply the upper *kuṭṭa* by the interpolator of the question and the lower *kuṭṭa* by the interpolator as reduced by the greatest common divisor. The same should be done reversely in the case of the fourth operation. In the case of these two operations, the *kuṭṭa* after being multiplied as indicated should be divided respectively by the dividend and divisor stated by the questioner, the residues will be the quotient and multiplier respectively.

"In the fifth operation, multiply the upper *kuṭṭa* by the greatest common divisor of the dividend and the interpolator, and the lower one by the other (*i.e.*, the greatest common divisor of the given divisor and the reduced interpolator). The products are the intermediate quotient and multiplier. Multiply the divisor of the question by the intermediate quotient and also its dividend by the intermediate multiplier. Difference of these products is the required intermediate divider. The intermediate quotient and multiplier are multiplied by the interpolator of the question and then divided by the intermediate divider. The quotients thus obtained being divided respectively by the dividend and divisor of the question, the residues will be the quotient and multiplier (required).

"The quotient and multiplier are obtained correctly by the process just described in the case of a positive interpolator when the chain is even and in the case of a negative interpolator if the chain is odd. In the case of an even chain and negative interpolator, also of an odd chain and positive interpolator, the quotient and multiplier thus obtained are subtracted respectively from the dividend and divisor made prime to each other and the residues give them correctly."[1]

The *rationale* of these rules will be easily found to be as follows:

(*i*) It will be noticed that to solve

$$by = ax \pm c, \tag{1}$$

in positive integers, Āryabhaṭa II first finds the solution of

$$by = ax \pm 1.$$

If $x = \alpha$, $y = \beta$ be a solution of this equation, we get

$$b\beta = a\alpha \pm 1,$$

or

$$b(c\beta) = a(c\alpha) \pm c.$$

Therefore $x = c\alpha$, $y = c\beta$ is a solution of (1).

(*ii*) Let $a = a'g$, $c = c'g$; then (1) reduces to

$$by' = a'x \pm c',$$

[1] *MSi*, xviii. 1-14.

where $y' = y/g$.

Let $x = \alpha$, $y' = \beta$ be a solution of

$$by' = a'x \pm 1,$$

so that we have

$$b\beta = a'\alpha \pm 1.$$

Hence $$bgc'\beta = a'gc'\alpha \pm c'g;$$

or $$b(c\beta) = a(c'\alpha) \pm c.$$

Therefore $x = c'\alpha$, $y = c\beta$ is a solution of (1).

(*iii*) Let $b = g'b'$, $c = g'c''$; then (1) reduces to

$$b'y = ax' \pm c'',$$

where $x' = x/g'$. If $x' = \alpha$, $y = \beta$ be a solution of

$$b'y = ax' \pm 1,$$

we have

$$b'\beta = a\alpha \pm 1.$$

Therefore $$b'g'c''\beta = ag'c''\alpha \pm g'c'',$$

or $$b(c''\beta) = a(c\alpha) \pm c.$$

Hence $x = c\alpha$, $y = c''\beta$ is a solution of (1).

(*iv*) Let $a = a'g$, $c = c'g$, $b = b''g''$ and $c' = c''g''$.

Then the given equation $by = ax \pm c$ reduces to

$$b''y' = a'x' \pm c'',$$

where $x' = x/g''$, $y' = y/g$. Now, if $x' = \alpha$, $y' = \beta$ be a solution of

$$b''y' = a'x' \pm 1,$$

we shall have, multiplying both sides by gg'',

$$b''gg''\beta = a'gg''\alpha \pm gg'',$$

or $$b(g\beta) = a(g''\alpha) \pm gg'',$$

or $$b\left\{\frac{c(g\beta)}{gg''}\right\} = a\left\{\frac{c(g''\alpha)}{gg''}\right\} \pm c.$$

Since $gg'' = a(g''\alpha) \sim b(g\beta)$, we get

$$b\left\{\frac{c(g\beta)}{a(g''\alpha) \sim b(g\beta)}\right\} = a\left\{\frac{c(g''\alpha)}{a(g''\alpha) \sim b(g\beta)}\right\} \pm c.$$

Therefore

$$x = \frac{c(g''\alpha)}{a(g''\alpha) \sim b(g\beta)}, \quad y = \frac{c(g\beta)}{a(g''\alpha) \sim b(g\beta)},$$

is a solution of the given equation $by = ax \pm c$. Since $c = c''gg'' = c''\{a(g''\alpha) \sim b(g\beta)\}$, both these values are integral.

In each of the above cases the minimum values of x, y satisfying the equation $by = ax \pm c$ are given by the residues left on dividing the values of x, y as calculated above by b and a respectively, provided the two quotients are equal.

Let $x = P$, $y = Q$ be the solution as calculated above; further suppose that

$$P = mb + p, \; Q = na + q;$$

where m, n are integers such that $p < b$, $q < a$.

If $m \neq n$, the minimum solution is either

$$\left.\begin{aligned} x &= p, \\ y &= (n - m)\,a + q \end{aligned}\right\}(1) \quad \text{or} \quad \left.\begin{aligned} x &= (m - n)b + p \\ y &= q \end{aligned}\right\}(2)$$

according as $m <$ or $> n$. Now, if the interpolator c is positive, it can be shown that (2) is not a solution. For, if it were,

$$\frac{bq - c}{a} = x, \quad \text{an integer,}$$

$$= (m - n)b + p > b.$$

But $q < a$, therefore,

$$\frac{bq - c}{a} < b,$$

which is absurd. Therefore, (1) must be the minimum solution in this case, not (2).

Similarly, if the interpolator c is negative, it can be shown that (2) is the minimum solution, not (1).

Hence the following rule of Āryabhaṭa II:

"If the quotients (m, n) obtained in the case of any proposed question be not equal, then the (derived) value for the multiplier should be accepted and that of the quotient rejected, if the interpolator is positive. On the other hand when the interpolator is negative, then the (derived) value for the quotient should be accepted and that for the multiplier rejected. How to obtain the quotient from the multiplier and the multiplier from the quotient correctly in all cases, I shall explain now. Multiply the (accepted) value of the multiplier by the dividend of the proposed question, add its interpolator and then divide by the divisor of the proposed question; the quotient is the corrected one. The product of the proposed divisor and the (accepted) quotient being added by the reverse of the interpolator and then divided by the dividend of the proposed question, the quotient is the (correct) multiplier."[1]

He has further indicated how to get all positive integral solutions of the equation $by = ax \pm c$ after having obtained the minimum solution.

"The (minimum) quotient and multiplier being added respectively with the dividend and divisor as stated in the question or as reduced, after multiplying both by an optional number, give various other values."[2]

That is to say, if $x = \alpha$, $y = \beta$ be the minimum solution, the general solution will be

$$x = bm + \alpha, \ y = am + \beta.$$

[1] *MSi*, xviii. 15-8. [2] *MSi*, xviii. 20.

Śrîpati's Rule. Śrîpati (1039) writes:

"Divide the dividend and divisor reciprocally until the residue is small. Set down the quotients one below the other in succession; then underneath them an optional number and below it the corresponding quotient, the optional number being determined thus: (the number) by which the last residue must be multiplied such that the product being subtracted by the interpolator and then divided by the divisor (corresponding to that residue), leaves no remainder. It is to be so when the number of quotients is even; in the case of an odd number of quotients the interpolator, if negative, must be first made positive and conversely, if positive, must be made negative; so it has been taught by the learned in this (branch of analysis). Now multiply the term above the optional number by it (the optional number) and then add the quotient below. Proceeding upwards such operation should be performed again and again until two numbers are obtained. The first one being divided by the divisor, (the residue) will give (the least value of) the multiplier; similarly the second being divided by the dividend, will give (the least value) of the quotient."[1]

Bhâskara II's Rules. Bhâskara II (1150) describes the method of the pulveriser thus:

"Divide mutually the dividend and divisor made prime to each other until unity becomes the remainder in the dividend. Set down the quotients one under the other successively; beneath them the interpolator and then cipher at the bottom. Multiply by the penultimate the number just above it and add the

[1] *SiŚe*, xiv. 22-25.
This rule is the same as that of Bhâskara I and holds under the same conditions. (See pp. 99f).

ultimate; then reject that ultimate. Do so repeatedly until only a pair of numbers is left. The upper one of these being divided by the reduced dividend, the remainder is the quotient; and the lower one being divided by the reduced divisor, the remainder is the multiplier. Such is precisely the process when the quotients (of mutual division) are even in number. But when they are odd, the quotient and multiplier so obtained must be subtracted from their respective abraders and the residues will be the true quotient and multiplier."[1]

Bhâskara II then shows how the process of solving a problem by the method of the pulveriser can sometimes be abbreviated to a great extent. He says:

"The multiplier is found by the method of the pulveriser after reducing the additive and dividend by their common divisor. Or, if the additive (previously reduced or not) and the divisor be so reduced, the multiplier found (by the method) being multiplied by their common measure will be the true one.

"Such is the process of finding the multiplier and quotient, when the interpolator is positive. On subtracting them from their respective abraders will be obtained the result for the subtractive interpolator."[2]

Kṛṣṇa (*c.* 1580) gives the following *rationale* of these rules:

We shall have to solve in positive integers

$$by = ax \pm c. \tag{1}$$

(*i*) Suppose g is the greatest common measure of a and c, so that $a = a'g$, $c = c'g$. Then

$$by = a'gx \pm c'g,$$

or

$$by' = a'x \pm c', \tag{1.1}$$

where $y' = y/g$. If $x = \alpha$, $y' = \beta$ be a solution of (1.1),

[1] *BBi*, pp. 25f; *L*, p. 77.

[2] *BBi*, p. 26; *L*, pp. 78, 79.

then clearly $x = \alpha$, $y = g\beta$ is a solution of (1).

(*ii*) Let $b = g'b'$, $c = g'c''$; then equation (1) reduces to

$$b'y = ax' \pm c'', \qquad (1.2)$$

where $x' = x/g'$. If $x' = \alpha'$, $y = \beta'$ be a solution of (1.2), then clearly $x = g'\alpha$, $y = \beta'$ is a solution of (1).

(*iii*) Let $a = a'g$, $c = c'g$; also $b = b''g''$, $c' = c''g''$; then equation (1) reduces to

$$b''y' = a'x' \pm c'', \qquad (1.3)$$

where $x' = x/g''$, $y' = y/g$. Then if $x' = \alpha$, $y' = \beta$ be a solution of (1.3), we shall have $x = g''\alpha$, $y = g\beta$ as a solution of (1).

Now, let the minimum solution of $by = ax + c$ be $x = \alpha$, $y = \beta$. Then

$$b\beta = a\alpha + c.$$

Hence $\qquad b(a - \beta) = a(b - \alpha) - c.$

Therefore, $\quad x = b - \alpha$, $y = a - \beta$ is a solution of $by = ax - c$. Since $\alpha < b$, $\beta < a$, provided $c < a, b$, this solution is positive. Thus we find that the minimum solution of the equation $by = ax - c$ can be derived from that of the equation $by = ax + c$, as has been stated by Bhâskara II.

Bhâskara II further observes :[1]

"In abrading the (calculated values of) the multiplier and the quotient (by the divisor and the dividend respectively) the intelligent should take out the same multiple (of them).

"The multiplier and quotient may be found as before after abrading the interpolator by the divisor; the quotient (obtained), however, must be increased by the abrading quotient in case the interpolator is positive, but, if it is negative, the abrading quotient

[1] *BBi*, p. 26; *L*, pp. 79, 81.

must be subtracted.

"Or the multiplier may be found as before after abrading both the dividend and the interpolator by the divisor; from (this multiplier) the quotient may be found by multiplying (it) by the dividend, adding (the interpolator) and then dividing (the sum by the divisor).[1]

"Those (minimum values of) the multiplier and the quotient being added by any (optionally chosen) multiple of their respective abraders become manifold."

We take the following illustrative example with the different methods of its solution from Bhâskara II:

To solve, in positive integers,

$$\frac{100x + 90}{63} = y.$$

First Method. Statement:

Dividend = 100
Divisor = 63   Additive = 90

Dividing mutually 100 by 63, we have

```
63) 100 (1
    63
    37) 63 (1
        37
        26) 37 (1
            26
            11) 26 (2
                22
                 4) 11 (2
                     8
                     3) 4 (1
                        3
                        1
```

[1] *i.e.*, by substituting the value of the multiplier in the original equation.

Then, forming the chain as directed in the rule, we get

1
1
1
2
2
1
90
0

By the rule, "Multiply by the penultimate the number just above it etc.," the two numbers obtained finally are 2430 and 1530.[1] Dividing these by 100 and 63 respectively, the remainders are 30 and 18. Hence $x = 18, y = 30$.

Second Method. Reducing the dividend and the additive by their greatest common divisor (10), we have the statement:

Dividend = 10
Divisor = 63   Additive = 9

Since

```
63) 10 (0
     0
    --
    10) 63 (6
        60
        --
         3) 10 (3
             9
            --
             1
```

[1] Successive operations in the application of the rule are :

1		1		1		1		1		~~1~~	2430
1		1		1		1		~~1~~	1530	~~1~~	1530
1		1		1		~~1~~	900	~~1~~	900	~~1~~	~~900~~
2		2		~~2~~	630	~~2~~	630	~~2~~	~~630~~	~~2~~	~~630~~
2		~~2~~	270	~~2~~	270	~~2~~	~~270~~	~~2~~	~~270~~	~~2~~	~~270~~
~~1~~	90	~~1~~	90	~~1~~	~~90~~	~~1~~	~~90~~	~~1~~	~~90~~	~~1~~	~~90~~
90		~~90~~		~~90~~		~~90~~		~~90~~		~~90~~	
~~0~~		~~0~~		~~0~~		~~0~~		~~0~~		~~0~~	

we get the chain

0
6
3
9
0

By the rule, "Multiply by the penultimate etc.," we obtain finally the numbers 27 and 171. Dividing them respectively by 10 and 63, we get the residues 7 and 45. Since the number of quotients of the mutual division is odd, subtracting 7 and 45 from the corresponding abraders 10 and 63, we get 3 and 18. In this case we neglect 3. So $x = 18$; whence by the given equation $y = 30$. Or, multiplying the quotient 3 as obtained above by the greatest common divisor 10, we get the same result $y = 30$.

Third Method. Reducing the divisor and the additive by their greatest common divisor (9), the statement is :

Dividend = 100
Divisor = 7
Additive = 10

Since

```
7) 100 (14
   98
   --
   2) 7 (3
      6
      -
      1
```

we get the chain

14
3
10
0

By the rule, "Multiply by the penultimate etc.," we obtain the two numbers 430 and 30. Dividing them by 100 and 7 respectively, the residues are 30 and 2.

Multiplying the latter by the greatest common divisor 9, we get $x = 18$ and $y = 30$.

Fourth Method. Dividing the divisor and the additive by their common measure (9) and again the dividend and the reduced additive by their common measure (10), we have

$$\text{Dividend} = 10 \qquad \text{Additive} = 1$$
$$\text{Divisor} = 7$$

Since

```
7) 10 (1
    7
   3) 7 (2
      6
      -
      1
```

we get the chain

$$\begin{array}{c} 1 \\ 2 \\ 1 \\ 0 \end{array}$$

By the rule, "Multiply by the penultimate etc.," we have finally the numbers 3 and 2. Dividing them by 10 and 7 respectively, the residues are the same. Multiplying them respectively by the common measure 10 of the dividend and reduced additive, and 9 of the divisor and additive, we get as before $x = 18$ and $y = 30$.

Adding to these minimum values (18, 30) of (x, y) optional multiples of the corresponding abraders (63, 100), we get the general solution of $100x + 90 = 63y$ in positive integers as $x = 63m + 18$, $y = 100m + 30$, where m is any integer.

Rules similar to those of Bhâskara II have been given by Nârâyaṇa,[1] Jñânarâja and Kamalâkara.[2]

[1] *NBi*, I, R. 55-60. [2] *SiTVi*, xiii. 183-190.

Solution of $by = ax \pm 1$

Constant Pulveriser. Though the simple indeterminate equation $by = ax \pm 1$ is solved exactly in the same way as the equation $by = ax \pm c$ and is indeed a particular case of the latter, yet on account of its special use in astronomical calculations[1] it has received separate consideration at the hands of most of the Hindu algebraists. It may, however, be noted that the separate treatment was somewhat necessitated by the physical conditions of the problems involving the two types. In the case of $by = ax \pm c$ the conditions are such that the value of either y or x, more particularly of the latter, has to be found and the rules for solution are formulated with that object. But in the case of the other ($by = ax \pm 1$) the physical conditions require the values of both y and x.

The equation $by = ax \pm 1$ is generally called by the name of *sthira-kuṭṭaka* or the "constant pulveriser" (from *sthira*, meaning constant, steady). Pṛthûdakasvâmî (860) sometimes designates it also as *dṛḍha-kuṭṭaka* (from *dṛḍha* = firm). But that name disappeared from later Hindu algebras because the word *dṛḍha* was employed by later writers[2] as equivalent to *niccheda* (having no divisor) or *nirapavarta* (irreducible). The origin of the name "constant pulveriser" has been explained by Pṛthûdakasvâmî as being due to the fact that the interpolator (± 1) is here invariable. Gaṇeśa[3] (1545) explains it in detail thus: In astronomical problems involving

[1] Thus Bhâskara II observes, "This method of calculation is of great use in mathematical astronomy." (*BBi*, p. 31). He then points out how the solutions of various astronomical problems can be derived from the solution of the same indeterminate equation. (*BBi*, p. 32; *L*, p. 81).

[2] This special technical use of the word *dṛḍha* occurs before Brahmagupta (628) in the works of Bhâskara I (522).

[3] *Vide* his commentary on the *Lîlâvatî* of Bhâskara II.

equations of the type $by - ax = \pm c$, the physical conditions are such that the dividend (a) and the divisor (b) are constant but the interpolator (c) always varies; so for their solution different sets of operations will have to be performed if we start directly to solve them all. But starting with the equation $by - ax = \pm 1$, we can derive the necessary solutions of all our equations from a constant set of operations. Hence the name is very significant. A similar explanation has been given by Kṛṣṇa (*c.* 1580).

Bhâskara I's Rule. Bhâskara I (522) writes:

"The method of the pulveriser is applied also after subtracting unity. The multiplier and quotient are respectively the numbers above and underneath. Multiplying those quantities by the desired number, divide by the reduced divisor and dividend; the residues are in this case known to be the (elapsed) days and (residues of) revolutions respectively."[1]

In other words, it has been stated that the solution of the equation

$$\frac{ax - c}{b} = y,$$

can be obtained by multiplying the solution of

$$\frac{ax - 1}{b} = y,$$

by c and then abrading as before. In general, the solution of the equation $by = ax \pm c$ in positive integers can be easily derived from that of $by = ax \pm 1$. If $x = \alpha$, $y = \beta$ be a solution of the latter equation, we shall have

$$b\beta = c\alpha \pm 1.$$

Then $$b(c\beta) = a(c\alpha) \pm c.$$

[1] *MBh*, i. 45.

Hence $x = c\alpha$, $y = c\beta$ is a solution of the former. The minimum solution will be obtained by abrading the values of x and y thus computed by b and a respectively, as indicated before.

Brahmagupta's Rule. To solve the equation $by = ax - 1$, Brahmagupta gives the following rule:

"Divide them (*i.e.*, the abraded coefficient of the multiplier and the divisor) mutually and set down the quotients one below the other. The last residue (of the reciprocal division after an even[1] number of quotients has been obtained) is multiplied by an optional integer such that the product being diminished by unity will be exactly divisible (by the divisor corresponding to that residue). The (optional) multiplier and then this quotient should be set down (underneath the listed quotients). Now proceeding from the lowermost term to the uppermost, by the penultimate multiply the term just above it and then add the lowermost number. (The uppermost number thus calculated) being divided by the reduced divisor, the residue (is the quantity required). This is the method of the constant pulveriser."[2]

Bhâskara II's Rule. Bhâskara II (1150) writes:

"The multiplier and quotient determined by supposing the additive or subtractive to be unity, multiplied severally by the desired additive or subtractive and then divided by their respective abraders, (the residues) will be those quantities corresponding to them (*i.e.*, desired interpolators)."[3]

This rule has been reproduced by Nârâyaṇa.[4] We take the following illustrative example with its solution

[1] In view of the rule in *BrSpSi*, xviii. 13.

[2] *BrSpSi*, xviii. 9-11.

[3] *BBi*, p. 31; *L*, p. 81.

[4] *NBi*, I, R. 65.

from Bhâskara II :[1]

$$\frac{221x + 65}{195} = y.$$

On dividing by the greatest common divisor 13, we get

$$\frac{17x + 5}{15} = y.$$

Now, by the method of the pulveriser the solution of the equation

$$\frac{17x + 1}{15} = y$$

is found to be $x=7, y=8$. Multiplying these values by 5 and then abrading by 15 and 17 respectively, we get the required minimum solution $x=5, y=6$.

Again a solution of

$$\frac{17x - 1}{15} = y$$

will be found to be $x=8$, $y=9$. Multiplying these quantities by 5 and abrading by 15 and 17, we get the solution of

$$\frac{17x - 5}{15} = y$$

to be $x=10$, $y=11$.

Solution of $by + ax = \pm c$

An equation of the form $by + ax = \pm c$ was generally transformed by Hindu algebraists into the form $by = - ax \pm c$ so that it appeared as a particular case of $by = ax \pm c$ in which a was negative.

Brahmagupta's Rule. Such an equation seems to

[1] *BBi*, pp. 28, 31; *L*, pp. 77, 81.

have been solved first by Brahmagupta (628). But his rule is rather obscure: "The reversal of the negative and positive should be made of the multiplier and interpolator."[1] Pṛthûdakasvâmî's explanation does not throw much light on it. He says, "If the multiplier be negative, it must be made positive; and the additive must be made negative: and then the method of the pulveriser should be employed." But he does not indicate how to derive the solution of the equation

$$by = -ax + c \qquad (1)$$

from that of the equation

$$by = ax - c \qquad (2)$$

The method, however, seems to have been this:

Let $x = \alpha$, $y = \beta$ be the minimum solution of (2). Then we get

$$b\beta = a\alpha - c$$

or

$$b(a - \beta) = -a(\alpha - b) + c.$$

Hence $x = \alpha - b$, $y = a - \beta$ is the minimum solution of (1). This has been expressly stated by Bhâskara II and others.

Bhâskara II's Rule. Bhâskara II says:

"Those (the multiplier and quotient) obtained for a positive dividend being treated in the same manner give the results corresponding to a negative dividend."[2]

The treatment alluded to in this rule is that of subtraction from the respective abraders. He has further elaborated it thus:

"The multiplier and quotient should be determined by taking the dividend, divisor and interpolator as positive. They will be the quantities for the additive interpolator. Subtracting them from their

[1] *BrSpSi*, xviii. 13. [2] *BBi*, p. 26.

respective abraders, the quantities for a negative interpolator are found. If the dividend or divisor be negative, the quotient should be stated as negative."[1]

Nârâyaṇa. Nârâyaṇa (1350) says:

"In the case of a negative dividend find the multiplier and quotient as in the case of its being positive and then subtract them from their respective abraders. One of these results, either the smaller one or the greater one, should be made negative and the other positive."[2]

Illustrative Examples. Examples with solutions from Bhâskara II:[3]

Example 1. $13y = -60x \pm 3.$

By the method described before we find that the minimum solution of

$$13y = 60x + 3$$

is $x = 11$, $y = 51$. Subtracting these values from their respective abraders, namely 13 and 60, we get 2 and 9. Then by the maxim. "In the case of the dividend and divisor being of different signs, the results from the operation of division should be known to be so," making the quotient negative we get the solution of

$$13y = -60x + 3$$

as $x = 2$, $y = -9$. Subtracting these values again from their respective abraders (13, 60), we get the solution of

$$13y = -60x - 3$$

as $x = 11$, $y = -51$.

Example 2. $-11y = 18x \pm 10.$

Proceeding as before we find the minimum solution of

$$11y = 18x + 10$$

[1] *BBi*, p. 29.

[2] *NBi*, I, R. 63.

[3] *BBi*, pp. 29, 30.

to be $x = 8$, $y = 14$. These will also be the values of x and y in the case of the negative divisor but the quotient for the reasons stated before should be made negative. So the solution of

$$-11y = 18x + 10$$

is $x = 8$, $y = -14$. Subtracting these (*i.e.*, their numerical values) from their respective abraders, we get the solution of

$$-11y = 18x - 10$$

as $x = 3$, $y = -4$.

"Whether the divisor is positive or negative, the numercial values of the quotient and multiplier remain the same: when either the divisor or the dividend is negative, the quotient must always be known to be negative."

The following example with its solution is from the algebra of Nârâyaṇa :[1]

$$7y = -30x \pm 3.$$

The solution of

$$7y = 30x + 3$$

is $x = 2$, $y = 9$. Subtracting these values from the respective abraders, namely 7 and 30, and making one of the remainders negative, we get $x = 5$, $y = -21$ and $x = -5$, $y = 21$ respectively as solutions of

$$7y = -30x \pm 3.$$

Particular Cases. The Hindus also found special types of general solutions of certain particular cases of the equation $by + ax = c$. For instance, we find in the *Gaṇita-sâra-saṁgraha* of Mahâvîra (850) problems of the following type :

"The *varṇa* (or colours) of two pieces of gold weighing 16 and 10 are unknown, but the mixture of

[1] *NBi*, I, Ex. 29.

them has the *varṇa* 4; what is the *varṇa* of each piece of gold?"[1]

If x, y denote the required *varṇa*, then we shall have

$$16x + 10y = 4 \times 26;$$

or in general

$$ax + by = c(a + b).$$

Therefore $\quad a(x - c) = b(c - y);$

whence $\quad x = c \pm m/a, \quad y = c \mp m/b,$

where m is an arbitrary integer.

Hence the following rule of Mahâvîra:

"Divide unity (severally) by the weights of the two ingots of gold. The resulting *varṇa* being set down at two places, increase or decrease it at one place and do reversely at the other place, by the unity divided by its own quantity of gold (the results will be the corresponding *varṇa*)."[2]

He has also remarked that "assuming an arbitrary value for one of the *varṇa*, the other can be found as before."[3]

A variation of the above problem is found in the *Lîlâvatî* of Bhâskara II:

"On mixing up two ingots of gold of *varṇa* 16 and 10 is produced gold of *varṇa* 12; tell me, O friend, the weights of the original ingots."[4]

That is to say, we shall have to solve the equation

$$16x + 10y = 12(x + y);$$

or in general

$$ax + by = c(x + y).$$

Hence $\quad x = m(c - b), \quad y = m(a - c),$

where m is an arbitrary integer.

[1] *GSS*, vi. 188.
[2] *GSS*, vi. 187.
[3] *GSS*, vi. 189.
[4] *L*, p. 26.

Hence the rule of Bhâskara II :

"Subtract the resulting *varṇa* from the higher *varṇa* and diminish it by the lower *varṇa*; the remainders multiplied by an optional number will be the weights of gold of the lower and higher *varṇa* respectively."[1]

In the above example $c - b = 2$, $a - c = 4$. So that, taking $m = 1$, 2, or 1/2, Bhâskara II obtains the values of (x, y) as (2, 4), (4, 8) or (1, 2). He then observes that in the same way numerous other sets of values can be obtained.

14. ONE LINEAR EQUATION IN MORE THAN TWO UNKNOWNS

To solve a linear equation involving more than two unknowns the usual Hindu method is to assume arbitrary values for all the unknowns except two and then to apply the method of the pulveriser. Thus Brahmagupta remarks, "The method of the pulveriser (should be employed), if there be present many unknowns (in an equation)."[2] Similar directions have been given by Bhâskara II and others.[3]

One of the astronomical problems proposed by Brahmagupta[4] leads to the equation :

$$197x - 1644y - z = 6302.$$

Hence $$x = \frac{1644y + z + 6302}{197}.$$

The commentator assumes $z = 131$. Then

$$x = \frac{1644y + 6433}{197};$$

[1] *L*, p. 25.
[2] *BrSpSi*, xviii. 51.
[3] *BBi*, p. 76.
[4] *BrSpSi*, xviii. 55.

hence by the method of the pulveriser

$$x = 41, \; y = 1.$$

The following example with its solution is from the algebra of Bhâskara II:

"The numbers of flawless rubies, sapphires, and pearls with one person are respectively 5, 8 and 7; and O friend, another has 7, 9 and 6 respectively of the same gems. In addition they have coins to the extent of 90 and 62. They are thus equally rich. Tell quickly, O intelligent algebraist, the price of each gem."[1]

If x, y, z represent the prices of a ruby, sapphire and pearl respectively, then by the question

$$5x + 8y + 7z + 90 = 7x + 9y + 6z + 62.$$

Therefore $\qquad x = \dfrac{-y + z + 28}{2}.$

Assume $z = 1$; then

$$x = \frac{-y + 29}{2};$$

whence by the method of the pulveriser, we get

$$x = 14 - m, \; y = 2m + 1,$$

where m is an arbitrary integer. Putting $m = 0, 1, 2, 3,\ldots$ we get the values of (x, y, z) as $(14, 1, 1)$, $(13, 3, 1)$, $(12, 5, 1)$, $(11, 7, 1)$, etc. Bhâskara II then observes, "By virtue of a variety of assumptions multiplicity of values may thus be obtained."

Sometimes the values of most of the unknowns present in an equation are assumed arbitrarily or in terms of any one of them, so as to reduce the equation to a simple determinate one. Thus Bhâskara II says:

"In case of two or more unknowns, x multiplied by 2 etc. (*i.e.*, by arbitrary known numbers), or divided,

[1] *BBi*, p. 77.

increased or decreased by them, or in some cases (simply) any known values may be assumed for the other unknowns according to one's own sagacity. Knowing these (the rest is an equation in one unknown)."[1]

The above example has been solved again by Bhâskara II in accordance with this rule thus :[2]

(1) Assume $x = 3z$, $y = 2z$. Then the equation reduces to

$$38z + 90 = 45z + 62.$$

Therefore $z = 4$. Hence $x = 12$, $y = 8$.

(2) Or assume $y = 5$, $z = 3$. Then the equation becomes

$$5x + 151 = 7x + 125.$$

Whence $x = 13$.

15. SIMULTANEOUS INDETERMINATE EQUATIONS OF THE FIRST DEGREE

Śrîpati's Rule. We have described before the rule of Brahmagupta for the solution of simultaneous equations of the first degree.[3] In the latter portion of that rule there are hints for the solution of simultaneous indeterminate equations by the application of the method of the pulveriser. Similar rules have been given by later Hindu algebraists. Thus Śrîpati (1039) says :

"Remove the first unknown from any one side of an equation leaving the rest, and remove the rest from the other side. Then find the value of the first by dividing the other side by its coefficient. If there be found thus several values (of the first unknown), the same (opera-

[1] *BBi*, p. 44.

[2] *BBi*, p. 46.

[3] See pp. 54f.

tions) should be made again (by equating two and two of those values) after reducing them to a common denominator. (Proceed thus repeatedly) until there results a single value for an unknown. Now apply the method of the pulveriser; and from the values (determined in this way) the other unknowns will be found by proceeding backwards. In the pulveriser the multiplier will be the value of the unknown associated with the dividend and the quotient, of that with the divisor."[1]

Bhâskara II's Rule. Bhâskara II (1150) writes:

"Remove the first unknown from the second side of an equation and the others as well as the absolute number from the first side. Then on dividing the second side by the coefficient of the first unknown, its value will be obtained. If there be found in this way several values of the same unknown, from them, after reduction to a common denominator and then dropping it, values of another unknown should be determined. In the final stage of this process, the multiplier and quotient obtained by the method of the pulveriser will be the values of the unknowns associated with the dividend and the divisor (respectively). If there be several unknowns in the dividend, their values should be determined after assuming values of all but one arbitrarily. Substituting these values and proceeding reversely, the values of the other unknowns can be obtained. If on so doing there results a fractional value (at any stage), the method of the pulveriser should be employed again. Then determining the (integral) values of the latter unknowns accordingly and substituting them, the values of the former unknowns should be found proceeding reversely again."[2]

A similar rule has been given by Jñânarâja.

[1] *SiŚe*, xiv. 15-6. [2] *BBi*, p. 76.

Example from Bhâskara II:

" (Four merchants), who have horses 5, 3, 6 and 8 respectively; camels 2, 7, 4 and 1; whose mules are 8, 2, 1 and 3; and oxen 7, 1, 2 and 1 in number; are all owners of equal wealth. Tell me instantly the price of a horse, etc."[1]

If x, y, z, w denote respectively the prices of a horse, a camel, a mule and an ox, and W be the total wealth of each merchant, we have

$$5x + 2y + 8z + 7w = W \quad (1)$$
$$3x + 7y + 2z + w = W \quad (2)$$
$$6x + 4y + z + 2w = W \quad (3)$$
$$8x + y + 3z + w = W \quad (4)$$

Then $x = \frac{1}{2}(5y - 6z - 6w)$, from (1) and (2)

$= \frac{1}{3}(3y + z - w)$, from (2) and (3)

$= \frac{1}{2}(3y - 2z + w)$, from (3) and (4)

From the first and second values of x, we get

$$y = \frac{1}{9}(20z + 16w);$$

and from the second and third values, we have

$$y = \frac{1}{3}(8z - 5w).$$

Equating these two values of y and simplifying,

$$20z + 16w = 24z - 15w.$$

Therefore $z = \dfrac{31w}{4}.$

Take $w = 4t$; then

$$z = 31t, \quad y = 76t, \quad x = 85t.$$

Special Rules. Bhâskara II observes that the physical conditions of problems may sometimes be such that the ordinary method of solving simultaneous in-

[1] *BBi*, p. 79.

9

determinate equations of the first degree, which has been just explained, will fail to give the desired result. One such problem has been described by him as follows:

"Tell quickly, O algebraist, what number is that which multiplied by 23 and severally divided by 60 and 80 leaves remainders whose sum is 100."[1]

Let the number be denoted by x; the quotients by u, v; and the remainders by s, t. Then we have

$$\frac{23x - s}{60} = u, \quad \frac{23x - t}{80} = v;$$

also $$s + t = 100.$$

Therefore $$x = \frac{60u + s}{23} = \frac{80v + t}{23}.$$

Hence $$x = \frac{60u + 80v + s + t}{46},$$

or $$x = \frac{30u + 40v + 50}{23}.$$

For the solution of the above he observes:

"Here, (although) there is more than one quotient (u, v) in the dividend, the value of any should not be arbitrarily assumed; for on so doing the process will fail."[2] "In a case like this," continues he, "the (given) sum of the remainders should be so broken up that each remainder will be less than the divisor corresponding to it and further that impossibility will not arise; then must be applied the usual method."

In the present example we thus suppose $s = 40$, $t = 60$. Hence we have

$$60u + 40 = 80v + 60$$

[1] *BBi*, p. 91. [2] *BBi*, p. 91f.

or $$u = \frac{80v + 20}{60} = \frac{4v + 1}{3};$$

whence by the method of the pulveriser, we get

$$v = 3w + 2, \ u = 4w + 3.$$

Therefore $$x = \frac{240w + 220}{23}.$$

Again, applying the method of the pulveriser in order to obtain an integral value of x, we have

$$w = 23m + 1, \quad x = 240m + 20.$$

If we take $s = 30$, $t = 70$, we shall find, by proceeding in the same way, another value of x as $240m + 90$.

General Problem of Remainders. One type of simultaneous indeterminate equations of the first degree is furnished by the general problem of remainders, *viz.*, to find a number N which being severally divided by $a_1, a_2, a_3, \ldots, a_n$ leaves as remainders $r_1, r_2, r_3, \ldots, r_n$ respectively.

In this case, we have the equations

$$N = a_1x_1 + r_1 = a_2x_2 + r_2 = a_3x_3 + r_3 = \ldots$$
$$= a_nx_n + r_n.$$

The method of solution of these equations was known to Āryabhaṭa I (499). For this purpose the term *dvicchedâgraṃ* occurring in his rule for the pulveriser must be explained in a different way so that the last line of the translations given before (pp. 94-5) will have to be replaced by the following: "(The result will be) the remainder corresponding to the product of the two divisors."[1] This explanation is, in fact, given by Bhâskara I, the direct disciple and earliest commentator of Āryabhaṭa I. Such a rule is expressly stated by

[1] See Bibhutibhusan Datta, *BCMS*, XXIV, 1932.

Brahmagupta.[1]

The *rationale* of this method is simple: Starting with the consideration of the first two divisors, we have

$$N = a_1x_1 + r_1 = a_2x_2 + r_2.$$

By the method described before we can find the minimum value α of x_1 satisfying this equation. Then the minimum value of N will be $a_1\alpha + r_1$. Hence the general value of N will be given by

$$N = a_1(a_2t + \alpha) + r_1,$$
$$= a_1a_2t + a_1\alpha + r_1,$$

where t is an integer. Thus $a_1\alpha + r_1$ is the remainder left on dividing N by a_1a_2, as stated by Āryabhaṭa I and Brahmagupta. Now, taking into consideration the third condition, we have

$$N = a_1a_2t + a_1\alpha + r_1 = a_3x_3 + r_3,$$

which can be solved in the same way as before. Proceeding in this way successively we shall ultimately arrive at a value of N satisfying all the conditions.

Pṛthûdakasvâmî remarks:

"Wherever the reduction of two divisors by a common measure is possible, there 'the product of the divisors' should be understood as equivalent to the product of the divisor corresponding to the greater remainder and quotient of the divisor corresponding to the smaller remainder as reduced (*i.e.*, divided) by the common measure.[2] When one divisor is exactly divisible by the other then the greater remainder is the (required) remainder and the divisor corresponding to

[1] *BrSpSi*, xviii. 5.

[2] *i.e.*, if p be the L.C.M. of a_1 and a_2, the general value of N satisfying the above two conditions will be

$$N = pt + a_1\alpha + r_1$$

instead of

$$N = a_1a_2t + a_1\alpha + r_1.$$

the greater remainder is taken as 'the product of the divisors.' (The truth of) this may be investigated by an intelligent mathematician by taking several symbols."

Examples from Bhâskara I :

(1) "Find that number which divided by 8 leaves 5 as remainder, divided by 9 leaves 4 as remainder and divided by 7 leaves 1 as remainder."[1]

That is to say, we have to solve

$$N = 8x + 5 = 9y + 4 = 7z + 1.$$

The solution is given substantially thus : The minimum value of N satisfying the first two conditions

$$N = 8x + 5 = 9y + 4$$

is found by the method of the pulveriser to be 13. This is the remainder left on dividing the number by the product 8.9. Hence

$$N = 72t + 13 = 7z + 1.$$

Again, applying the same method we find the minimum number satisfying all the conditions to be 85.

(2) "Tell me at once, O mathematician, that number which leaves unity as remainder when divided by any of the numbers from 2 to 6 but is exactly divisible by 7."

By the same method, says Bhâskara I (522), the number is found to be 721. By a different method Sûryadeva Yajvâ obtains the number 301. It is interesting to find that this very problem was afterwards treated by Ibn-al-Haitam (*c.* 1000) and Leonardo Fibonacci of Pisa (*c.* 1202).[2]

To solve a problem of this kind Bhâskara II adopts

[1] See his commentary on *A*, ii. 32-3.

[2] L.E. Dickson, *History of the theory of Numbers*, Vol. II, referred to hereafter as Dickson, *Numbers* II, pp. 59, 60.

two methods. One is identical with the method of Āryabhaṭa I and the other follows from his general rule for the solution of simultaneous indeterminate equations of the first degree. They will be better understood from his applications to the solution[1] of the following problem which, as Pṛthûdakasvâmî (860) observes,[2] was popular amongst the Hindus:

To find a number N which leaves remainders 5, 4, 3, 2 when divided by 6, 5, 4, 3 respectively.

i.e., $N = 6x + 5 = 5y + 4 = 4z + 3 = 3w + 2.$

(1) We have

$$x = \frac{5y - 1}{6}, \quad y = \frac{4z - 1}{5}, \quad z = \frac{3w - 1}{4}.$$

Now by the method of the pulveriser, we get from the last equation

$$w = 4t + 3, \quad z = 3t + 2,$$

where t is an arbitrary integer. Substituting in the second equation, we get

$$y = \frac{12t + 7}{5}.$$

To make this integral, we again apply the method of the pulveriser, so that

$$t = 5s + 4, \quad y = 12s + 11.$$

This value of y makes x a whole number. Hence we have finally

$$w = 20s + 19, z = 15s + 14, y = 12s + 11, x = 10s + 9.$$

$$\therefore \quad N = 60s + 59.$$

(2) Or we may proceed thus:

Since $\qquad N = 6x + 5 = 5y + 4,$

[1] *BBi*, pp. 85f.

[2] *Vide* his commentary on *BrSpSi*, xviii. 3-6.

we have $$x = \frac{5y - 1}{6}.$$

But x must be integral, so $y = 6t + 5$, $x = 5t + 4$.

Hence $$N = 30t + 29.$$

Again $$N = 30t + 29 = 4z + 3.$$

$$\therefore \qquad t = \frac{2z - 13}{15}.$$

Since t must be integral, we must have $z = 15s + 14$; hence $t = 2s + 1$. Therefore

$$N = 60s + 59.$$

The last condition is identically satisfied. Pṛthûdakasvâmî followed this second method to solve the above problem.

Conjunct Pulveriser. The foregoing system of indeterminate equations of the first degree can be put into the form[1]

$$\left.\begin{array}{l} by_1 = a_1x \pm c_1 \\ by_2 = a_2x \pm c_2 \\ by_3 = a_3x \pm c_3 \\ \dots\dots\dots\dots\dots\dots \end{array}\right\} \qquad (1)$$

On account of its important applications in mathematical astronomy this modified system has received special treatment at the hands of Hindu algebraists from Âryabhaṭa II (950) onwards. It is technically called

[1] For, we have

$$a_1x_1 + r_1 = a_2x_2 + r_2 = a_3x_3 + r_3 = \dots = a_nx_n + r_n.$$

Then
$$a_2x_2 = a_1x_1 + (r_1 - r_2),$$
$$a_2x_3 = \frac{a_1a_2}{a_3}x_1 + \frac{a_2}{a_3}(r_1 - r_3),$$
$$a_2x_4 = \frac{a_1a_2}{a_4}x_1 + \frac{a_2}{a_4}(r_1 - r_4),$$
$$\dots\dots\dots\dots\dots\dots\dots\dots\dots\dots$$

saṁśliṣṭakuṭṭaka or the "conjunct pulveriser" (from, *kuṭṭaka* = pulveriser and *saṁśliṣṭa* = joined together, related).

For the solution of the above system of equations Āryabhaṭa II lays down the following rule :

"In the solution of simultaneous indeterminate equations of the first degree with a common divisor, the dividend will be the sum of the multipliers[1] and the interpolator the sum of the given interpolators."[2]

A similar rule is given by Bhâskara II. He says :

"If the divisor be the same but the multipliers different then making the sum of the multipliers the dividend and the sum of residues the residue (of a pulveriser), the investigation is carried on according to the foregoing method. This true method of the pulveriser is called the conjunct pulveriser."[3]

Rationale. If the equations (1) are satisfied by some value a of x, then the same value will satisfy the equation

$$b(y_1 + y_2 + \dots) = (a_1 + a_2 + \dots)x + (c_1 + c_2 + \dots) \quad (2).$$

Thus, if we can find the general value of x satisfying equation (2), one of these values, at least, will satisfy all the equations (1).

To illustrate the application of the above Bhâskara II gives the following example :[4]

$$\left.\begin{array}{l} 63y_1 = 5x - 7 \\ 63y_2 = 10x - 14 \end{array}\right\} \quad (A)$$

Adding up the equations and dividing by the common factor 3, we get

$$21Y = 5x - 7,$$

[1] In the equations (1), a_1, a_2, ... are called multipliers.

[2] *MSi*, xviii. 48.

[3] *BBi*, p. 33 ; *L*, p. 82.

[4] *BBi*, p. 33 ; *L*, p. 82.

where $Y = y_1 + y_2$. By the method of the pulveriser the least positive value of x satisfying this equation is $x = 14$. This value of x is found to satisfy both the equations (*A*).

Generalised Conjunct Pulveriser. A generalised case of the conjunct pulveriser is that in which the divisors as well as the multipliers vary. Thus we have

$$b_1y_1 = a_1x \pm c_1,$$
$$b_2y_2 = a_2x \pm c_2,$$
$$b_3y_3 = a_3x \pm c_3,$$
$$\cdots\cdots\cdots\cdots\cdots\cdots$$

Simultaneous indeterminate equations of this type have been treated by Mahâvîra (850) and Śrîpati (1039). Mahâvîra says:

"Find the least solutions of the first two equations. Divide the divisor corresponding to the greater solution by the other divisor (and as in the method of the pulveriser find the least value of) the multiplier with the difference of the solutions as the additive. That multiplied by the divisor (corresponding to the greater solution) and then added by the greater solution (will be the value of the unknown satisfying the two equations)."[1]

A similar rule is given by Śrîpati:

"Find the least solutions of the first two equations. Dividing the divisor corresponding to the greater solution by the divisor corresponding to the smaller solution, the residue (and its divisor) should be mutually divided. Then taking the difference of the numbers as the additive, determine (the least value of) the multiplier of the divisor corresponding to the greater solution in the manner explained before. Multiply that value by the

[1] *GSS*, vi. 115½, 136½ (last lines).

latter divisor and then add the solution (corresponding to it). The resulting number (severally) multiplied by the two multipliers and divided by the corresponding divisors will leave remainders as stated."[1]

The *rationale* of these rules will be clear from the following:

Taking the first two equations, we have

$$b_1y_1 = a_1x \pm c_1,$$
$$b_2y_2 = a_2x \pm c_2.$$

Suppose α_1 to be the least value of x satisfying the first equation as found by the method of the pulveriser. Then $b_1m + \alpha_1$, where m is an arbitrary integer, will be the general value of x satisfying that equation. Similarly, we shall find from the second equation the general value of x as $b_2n + \alpha_2$. If the same value of x satisfies both the equations we must have

$$b_2n + \alpha_2 = b_1m + \alpha_1,$$

or $$b_2n = b_1m + (\alpha_1 - \alpha_2);$$

supposing $\alpha_1 > \alpha_2$. Solving this equation, we can find the value of m and hence of $b_1m + \alpha_1$ of x satisfying both the equations. The general value of x derived from this may be equated to the value of x from the third equation and the resulting equation solved again, and so on.

In illustration of his rule Mahâvîra proposed several problems. One of these has already been given (Part I, p. 233). Here are two others:

(1) "Five (heaps of fruits) added with two (fruits) were divided (equally) between nine travellers; six (heaps) added with four (fruits) were divided amongst eight; four (heaps) increased by one (fruit) were divided

[1] *SiŚe*, xiv. 28.

amongst seven. Tell the number (of fruits in each heap)."[1]

This gives the equations:

$$9y_1 = 5x + 2, \quad 8y_2 = 6x + 4, \quad 7y_3 = 4x + 1.$$

(2) "The (dividends) are the sixteen numbers beginning with 35 and increasing successively by three; divisors are 32 and others successively increasing by 2; and 1 increasing by 3 gives the remainders positive and negative. What is the unknown multiplier?"[2]

This gives the equations:

$$32y_1 = 35x \pm 1, \quad 34y_2 = 38x \pm 4, \quad 36y_3 = 41x \pm 7, \ldots$$

Alternative Method. In four palm-leaf manuscript copies of the *Lîlâvatî* of Bhâskara II Sarada Kanta Ganguly discovered a rule describing an alternative method for the solution of the generalised conjunct pulveriser.[3] There is also an illustrative example. The genuineness of this rule and example is accepted by him; but it has been questioned by A. A. Krishnaswami Ayyangar[4] who attributes them to some commentator of the work. His arguments are not convincing.[5] The chief points against the presumption, which have been noted also by Ganguly, are: (1) the rule and example in question have not been mentioned by the earlier commentators of the *Lîlâvatî* and (2) they have not been so far traced in any manuscript of the *Bîjagaṇita*, though the treatment of the pulveriser occurs nearly word for

[1] *GSS*, vi. 129½. [2] *GSS*, vi. 138½.

[3] S. K. Ganguly, "Bhâskarâcârya and simultaneous indeterminate equations of the first degree," *BCMS*, XVII, 1926, pp. 89-98.

[4] A. A. Krishnaswami Ayyangar, "Bhâskara and samslishta Kuttaka," *JIMS*, XVIII, 1929.

[5] For Ganguly's reply to Ayyangar's criticism see *JIMS*, XIX, 1931.

word in the two works. Still we are in favour of accepting Ganguly's conclusion.[1] The rule in question is this:

"If the divisors as well as the multipliers be different, find the value of the unknown answering to the first set of them. That value being multiplied by the second dividend and then added by the second interpolator will be the interpolator (of a new *kuṭṭaka*); the product of the second dividend and first divisor will be the dividend there and the divisor will be the second divisor. The value of the unknown multiplier determined from the *kuṭṭaka* thus formed being multiplied by the first divisor and added by the previous value of the unknown multiplier will be the value (answering to the two divisors). The dividend (for the next step) has been stated to be equal to the product of the two divisors. So proceed in the same way with the third divisor. And so on with the others, if there be many."

The *rationale* of this rule is as follows: Let α_1 be the least value of x satisfying the first equation of the system, *viz.*,

$$b_1 y_1 = a_1 x \pm c_1.$$

Hence the general value is $x = b_1 t + \alpha_1$, where t is any integer. Substituting this value in the second equation, we get

$$b_2 y_2 = a_2 b_1 t + (a_2 \alpha_1 \pm c_2).$$

If $t = \tau$ be a solution of this equation, a value of x

[1] Of the four manuscripts containing the rule and example in question two are from Puri, in Oriya characters, with the commentary of Śrîdhara Mahâpâtra (1717); the other two, in Andhra characters and without any commentary, are preserved in the Oriental Libraries of Madras and Mysore. So these four manuscript copies do not appear to have been drawn from the same source. This is a strong point in favour of the genuineness of the rule and example.

satisfying both the equations will be $\alpha_2 = b_1\tau + \alpha_1$ as stated in the rule. Now the general value of t will be $t = b_2m + \tau$, where m is an integer. Hence $x = b_1t + \alpha_1 = b_1b_2m + b_1\tau + \alpha_1 = b_1b_2m + \alpha_2$. Substituting this value in the third equation we can find the least value of m and hence a value of x answering to the three equations. And so on for the other equations.

The example runs thus :

"Tell me that number which multiplied by 7 and then divided by 62, leaves the remainder 3. That number again when multiplied by 6 and divided by 101 leaves the remainder 5 ; and when multiplied by 8 and divided by 17 leaves the remainder 9. Also (give) at once the process of the pulveriser for (finding) the number with the remainders all positive."

Symbolically, we have

(1) $62y_1 = 7x - 3,\ 101y_2 = 6x - 5,\ 17y_3 = 8x - 9;$

(2) $62y_1 = 7x + 3,\ 101y_2 = 6x + 5,\ 17y_3 = 8x + 9.$

16. SOLUTION OF $Nx^2 + 1 = y^2$

Square-nature. The indeterminate quadratic equation

$$Nx^2 \pm c = y^2,$$

is called by the Hindus *Varga-prakṛti* or *Kṛti-prakṛti*, meaning the "Square-nature."[1] Bhâskara II (1150) states that the absolute number should be *rûpa*,[2] which means "unity" as well as "absolute number" in general. Kamalâkara (1658) says :

[1] *Varga* = *kṛti* = "square" and *prakṛti* = "nature," "principle," "origin," etc. Colebrooke has rendered the term *varga-prakṛti* as "Affected Square."

[2] "Tatra rûpakṣepapadârthaṁ tâvat"—*BBi*, p. 33.

"Hear first the nature of the *varga-prakṛti* : in it the square (of a certain number) multiplied by a multiplier and then increased or diminished by an interpolator becomes capable of yielding a square-root."[1]

It was recognised that the most fundamental equation of this class is

$$Nx^2 + 1 = y^2,$$

where N is a non-square integer.

Origin of the Name. As regards the origin of the name *varga-prakṛti*, Kṛṣṇa (1580) says: "That in which the *varga* (square) is the *prakṛti* (nature) is called the *varga-prakṛti* ; for the square of *yâvat*, etc., is the *prakṛti* (origin) of this (branch of) mathematics. Or, because this (branch of) mathematics has originated from the number which is the *prakṛti* of the square of *yâvat*, etc., so it is called the *varga-prakṛti*. In this case the number which is the multiplier of the square of *yâvat*, etc., is denoted by the term *prakṛti*. (In other words) it is the coefficient of the square of the unknown."[2] This double interpretation has been evidently suggested by the use of the term *prakṛti* by Bhâskara II in two contexts. He has denoted by it sometimes the quantity N of the above equation as in "There the number which is (associated) with the square of the unknown is the *prakṛti*;"[3] and at other times x^2, as in "Supposing the square of one of the two unknowns to be the *prakṛti*."[4] Other Hindu algebraists have, however, consistently

[1] *SiTVi*, xiii. 208.

[2] See his commentary on the *Bîjagaṇita* of Bhâskara II.

[3] "Tatra varṇavarge yo'ṅkaḥ sâ prakṛtiḥ" (*BBi*, p. 100). Compare also "Tatra yâvattâvadvarge yo'ṅkaḥ sâ prakṛtiḥ" (p. 107) ; "Iṣṭaṁ hrasvaṁ tasya vargaḥ prakṛtyâ kṣuṇṇo..." (p. 33).

[4] "Tatraikâṁ varṇakṛtiṁ prakṛtiṁ prakalpya..." (*BBi*, p. 106). Compare also "Sarûpake varṇakṛti tu yatra tatrecchaikâṁ prakṛtiṁ prakalpya..." (p. 105).

employed the term *prakṛti* to denote N only.[1] Brahmagupta (628) uses the term *guṇaka* (multiplier) for the same purpose.[2] This latter term, together with its variation *guṇa*, appears occasionally also in later works.[3]

We presume that the name *varga-prakṛti* originated from the following consideration: The principle (*prakṛti*) underlying the calculations in this branch of mathematics is to determine a number (or numbers) whose nature (*prakṛti*) is such that its (or their) square (or squares, *varga*) or the simple number (or numbers) after certain specified operations will yield another number (or numbers) of the nature of a square. So the name is, indeed, very significant. This interpretation seems to have been intended, at any rate, by the earlier writers who used the term in a wider sense.[4] It is perhaps noteworthy that we do not find in the works of Brahmagupta the use of the word *prakṛti* either in the sense of N or of x^2.

Technical Terms. Of the various technical terms which are ordinarily used by the Hindu algebraists in connection with the Square-nature we have already dealt with the most notable one, *prakṛti*, together with its synonyms. Others have been explained by Pṛthûdakasvâmî (860) thus:

"Here are stated for ordinary use the terms which

[1] For instance, Pṛthûdakasvâmî (860) writes: "The multiplier (of the square of the unknown) is known as the *prakṛti*;" Śrîpati (1039): "Kṛter-guṇako prakṛtirbhṛśoktaḥ" (*SiŚe*, xiv. 32); Kamalâkara: "Guṇo yo râśi-vargasya saiva prakṛtirucyate."

[2] *BrSpSi*, xviii. 64.

[3] For instance, Śrîpati employs the term *guṇaka* (*SiŚe*, xiv. 32); Bhâskara II and Nârâyaṇa use *guṇa* (*BBi*, p. 42; *NBi*, I, R. 84).

[4] For instance, Brahmagupta seems to have considered the scope of the subject wide enough to include such equations as

$$x + y = u^2,\ x - y = v^2,\ xy + 1 = w^2,$$

amongst others (cf. *BrSpSi*, xviii. 72).

are well known to people. The number whose square, multiplied by an optional multiplier and then increased or decreased by another optional number, becomes capable of yielding a square-root, is designated by the term the lesser root (*kaniṣṭha-pada*) or the first root (*âdya-mûla*). The root which results, after those operations have been performed, is called by the name the greater root (*jyeṣṭha-pada*) or the second root (*anya-mûla*). If there be a number multiplying both these roots, it is called the augmenter (*udvartaka*) ; and, on the contrary, if there be a number dividing the roots, it is called the abridger (*apavartaka*)."[1]

Bhâskara II (1150) writes :

"An optionally chosen number is taken as the lesser root (*hrasva-mûla*). That number, positive or negative, which being added to or subtracted from its square multiplied by the *prakṛti* (multiplier) gives a result yielding a square-root, is called the interpolator (*kṣepaka*). And this (resulting) root is called the greater root (*jyeṣṭha-mûla*)."[2]

Similar passages occur in the works of Nârâyaṇa,[3] Jñânarâja and Kamalâkara.[4]

The terms 'lesser root' and 'greater root' do not appear to be accurate and happy. For if $x = m, y = n$ be a solution of the equation $Nx^2 + c = y^2$, m will be less than n, if N and c are both positive. But if they are of opposite signs, the reverse will sometimes happen.[5]

[1] See Pṛthûdakasvâmî's commentary on *BrSpSi*, xviii. 64. In the equation $Nx^2 \pm c = y^2$, $x =$ lesser root, $y =$ greater root, $N =$ multiplier, and $c =$ interpolator.

[2] *BBi*, p. 33.

[3] *NBi*, I, R. 72.

[4] *SiTVi*, xiii. 209.

[5] For instance, take the following example from Bhâskara II (*BBi*, p. 43) :

Therefore, in the latter case, where $m > n$, it will be obviously ambiguous to call m the lesser root and n the greater root, as was the practice in later Hindu algebra. This defect in the prevalent terminology was noticed by Kṛṣṇa (1580). He explains it thus: "These terms are significant. Where the greater root is sometimes smaller than the lesser root owing to the interpolator being negative, there also it becomes greater than the lesser root after the application of the Principle of Composition."[1] The earlier terms, 'the first root' (*âdya-mûla*) for the value of x and 'the second root' or 'the last root' (*antya-mûla*) for the value of y, are quite free from ambiguity. Their use is found in the algebra of Brahmagupta (628).[2] The later terms appear in the works of his commentator Pṛthûdakasvâmî (860).

The interpolator is called by Brahmagupta *kṣepa*, *prakṣepa* or *prakṣepaka*.[3] Śrîpati occasionally employs the synonym *kṣipti*.[4] When negative, the interpolator is sometimes distinguished as 'the subtractive' (*śodhaka*).

$$13x^2 - 13 = y^2.$$

One solution of it is given by the author as $x = 1, y = 0$; so that here the lesser root is greater than the greater root. The same is the case in the solution $x = 2, y = 1$ of his example (*BBi*, p. 43)

$$-5x^2 + 21 = y^2.$$

Brahmagupta gives the example (*BrSpSi*, xviii. 77)

$$3x^2 - 800 = y^2,$$

which has a solution ($x = 20, y = 20$) where the two roots are equal.

[1] For example, by composition of the solution (1, 0) of the equation $13x^2 - 13 = y^2$ with the solution ($\frac{3}{2}$, $\frac{11}{2}$) of the equation $13x^2 + 1 = y^2$, we obtain, after Bhâskara II, a new solution ($\frac{11}{2}$, $\frac{39}{2}$) of the former, in which the greater root is greater than the lesser root. Similarly, by composition of the solution (2, 1) of the equation $-5x^2 + 21 = y^2$ with the solution ($\frac{1}{3}$, $\frac{2}{3}$) of the equation $-5x^2 + 1 = y^2$, we get a new solution (1, 4) of the former satisfying the same condition.

[2] *BrSpSi*, xviii. 64, 66f.

[3] *BrSpSi*, xviii. 65.

[4] *SiŚe*, xiv. 32.

The positive interpolator is then called 'the additive.'[1]

Brahmagupta's Lemmas. Before proceeding to the general solution of the Square-nature Brahmagupta has established two important lemmas. He says:

"Of the square of an optional number multiplied by the *guṇaka* and increased or decreased by another optional number, (extract) the square-root. (Proceed) twice. The product of the first roots multiplied by the *guṇaka* together with the product of the second roots will give a (fresh) second root; the sum of their cross-products will be a (fresh) first root. The (corresponding) interpolator will be equal to the product of the (previous) interpolators."[2]

The rule is somewhat cryptic because the word *dvidhâ* (twice) has been employed with double implication. According to one, the earlier operations of finding roots are made on two optional numbers with two optional interpolators, and with the results thus obtained the subsequent operations of their composition are performed. According to the other implication of the word, the earlier operations are made with one optionally chosen number and one interpolator, and the subsequent ones are carried out after the repeated statement of those roots for the second time. It is also implied that in the composition of the quadratic roots their products may be added together or subtracted from each other.

That is to say, if $x = \alpha$, $y = \beta$ be a solution of the equation

$$Nx^2 + k = y^2,$$

and $x = \alpha'$, $y = \beta'$ be a solution of

$$Nx^2 + k' = y^2,$$

then, according to the above,

[1] *BrSpSi*, xviii. 64-5.

[2] *BrSpSi*, xviii. 64-5.

$$x = \alpha\beta' \pm \alpha'\beta,\ y = \beta\beta' \pm N\alpha\alpha'$$

is a solution of the equation

$$Nx^2 + kk' = y^2.$$

In other words, if

$$N\alpha^2 + k = \beta^2,$$
$$N\alpha'^2 + k' = \beta'^2,$$

then

$$N(\alpha\beta' \pm \alpha'\beta)^2 + kk' = (\beta\beta' \pm N\alpha\alpha')^2. \qquad \text{(I)}$$

In particular, taking $\alpha = \alpha'$, $\beta = \beta'$, and $k = k'$, Brahmagupta finds from a solution $x = \alpha$, $y = \beta$ of the equation

$$Nx^2 + k = y^2,$$

a solution $x = 2\alpha\beta$, $y = \beta^2 + N\alpha^2$ of the equation

$$Nx^2 + k^2 = y^2.$$

That is, if

$$N\alpha^2 + k = \beta^2,$$

then

$$N(2\alpha\beta)^2 + k^2 = (\beta^2 + N\alpha^2)^2. \qquad \text{(II)}$$

This result will be hereafter called *Brahmagupta's Corollary*.

Description by Later Writers. Brahmagupta's Lemmas have been described by Bhâskara II (1150) thus:

"Set down successively the lesser root, greater root and interpolator; and below them should be set down in order the same or another (set of similar quantities). From them by the Principle of Composition can be obtained numerous roots. Therefore, the Principle of Composition will be explained here. (Find) the two cross-products of the two lesser and the two greater roots; their sum is a lesser root. Add the product of the two lesser roots multiplied by the *prakṛti* to the product of the two greater roots; the sum will be a greater root. In that (equation) the interpolator will be the product of the two previous interpolators. Again the difference of the two cross-products is a lesser root. Subtract the product of the two lesser roots multiplied by the *prakṛti* from the product of the two greater roots; (the difference) will be a greater root. Here also, the interpolator is the product of the two (previous) interpolators."[1]

Statements similar to the above are found in the works of Nârâyaṇa[2] (1350), Jñanarâja (1503) and Kamalâkara[3] (1658).

Principle of Composition. The above results are called by the technical name, *Bhâvanâ* (demonstration or proof, meaning anything demonstrated or proved, hence theorem, lemma ; the word also means composition or combination). They are further distinguished as *Samâsa Bhâvanâ* (Addition Lemma or Additive Composition) and *Antara Bhâvanâ* (Subtraction Lemma or Subtractive Composition). Again, when the *Bhâvanâ* is made with two equal sets of roots and interpolators, it is called *Tulya Bhâvanâ* (Composition of Equals) and when with two unequal sets of values, *Atulya Bhâvanâ* (Composition of Unequals). Kṛṣṇa has observed that when it is desired to derive roots of a Square-nature, larger in value, one should have recourse to the Addition Lemma and for smaller roots one should use the Subtraction Lemma.

Brahmagupta's Lemmas were rediscovered and recognised as important by Euler in 1764 and by Lagrange in 1768.

Proof. The proof of Brahmagupta's Lemmas has been given by Kṛṣṇa substantially as follows :

[1] *BBi*, p. 34.
[2] *NBi*, I, R. 72-75½.
[3] *SiTVi*, xiii. 210-214.

We have

$$N\alpha^2 + k = \beta^2,$$
$$N\alpha'^2 + k' = \beta'^2.$$

Multiplying the first equation by β'^2, we get

$$N\alpha^2\beta'^2 + k\beta'^2 = \beta^2\beta'^2.$$

Now, substituting the value of the factor β'^2 of the interpolator from the second equation, we get

$$N\alpha^2\beta'^2 + k(N\alpha'^2 + k') = \beta^2\beta'^2,$$

or
$$N\alpha^2\beta'^2 + Nk\alpha'^2 + kk' = \beta^2\beta'^2.$$

Again, substituting the value of k from the first equation in the second term of the left-hand side expression, we have

$$N\alpha^2\beta'^2 + N\alpha'^2(\beta^2 - N\alpha^2) + kk' = \beta^2\beta'^2,$$

or
$$N(\alpha^2\beta'^2 + \alpha'^2\beta^2) + kk' = \beta^2\beta'^2 + N^2\alpha^2\alpha'^2.$$

Adding $\pm 2N\alpha\beta\alpha'\beta'$ to both sides, we get

$$N(\alpha\beta' \pm \alpha'\beta)^2 + kk' = (\beta\beta' \pm N\alpha\alpha')^2.$$

Brahmagupta's Corollary follows at once from the above by putting $\alpha' = \alpha$, $\beta' = \beta$ and $k' = k$.

General Solution of the Square-Nature. It is clear from Brahmagupta's Lemma (I) that when two solutions of the Square-nature,

$$Nx^2 + 1 = y^2,$$

are known, any number of other solutions can be found. For, if the two solutions be (a, b) and (a', b'), then two other solutions will be

$$x = ab' \pm a'b, \quad y = bb' \pm Naa'.$$

Again, composing this solution with the previous ones, we shall get other solutions. Further, it follows from Brahmagupta's Corollary that if (a, b) be a solution of the equation, another solution of it is $(2ab,\ b^2 + Na^2)$. Hence, in order to obtain a set of solutions of the

Square-nature it is necessary to obtain only one solution of it. For, after having obtained that, an infinite number of other solutions can be found by the repeated application of the Principle of Composition. Thus Śrîpati (1039) observes: "There will be an infinite (set of two roots)."[1] Bhâskara II (1150) remarks: "Here (*i.e.*, in the solution of the Square-nature) the roots are infinite by virtue of (the infinitely repeated application of) the Principle of Composition as well as of (the infinite variety of) the optional values (of the first roots)."[2] Nârâyaṇa (1350) writes, "By the Principle of Composition of equal as well as unequal sets of roots, (will be obtained) an infinite number of roots."[3]

Modern historians of mathematics are incorrect in stating that Fermat (1657) was the first to assert that the equation $Nx^2 + 1 = y^2$, where N is a non-square integer, has an unlimited number of solutions in integers.[4] The existence of an infinite number of integral solutions was clearly mentioned by Hindu algebraists long before Fermat.

Another Lemma. Brahmagupta says:

"On dividing the two roots (of a Square-nature) by the square-root of its additive or subtractive, the roots for the interpolator unity (will be found)."[5]

That is to say, if $x = \alpha$, $y = \beta$ be a solution of the equation

$$Nx^2 + k^2 = y^2,$$

then $x = \alpha/k$, $y = \beta/k$ is a solution of the equation

$$Nx^2 + 1 = y^2.$$

This rule has been restated in a different way thus:

[1] *SiSe*, xiv. 33.
[2] "Ihânantyaṁ bhâvanâbhistatheṣṭataḥ"—*BBi*, p. 34.
[3] *NBi*, I, R. 78. Compare also *SiTVi*, xiii. 217.
[4] Smith, *History*, II, p. 453.
[5] *BrSpSi*, xviii. 65.

"If the interpolator is that divided by a square then the roots will be those multiplied by its square-root."[1]

That is, suppose the Square-nature to be

$$Nx^2 \pm p^2d = y^2,$$

so that its interpolator p^2d is exactly divisible by the square p^2. Then, putting therein $u = x/p$, $v = y/p$, we derive the equation

$$Nu^2 \pm d = v^2,$$

whose interpolator is equal to that of the original Square-nature divided by p^2. It is clear that the roots of the original equation are p times those of the derived equation.

Bhâskara II writes:

"If the interpolator (of a Square-nature) divided by the square of an optional number be the interpolator (of another Square-nature), then the two roots (of the former) divided by that optional number will be the roots (of the other). Or, if the interpolator be multiplied, the roots should be multiplied."[2]

The same rule has been stated in slightly different words by Nârâyaṇa[3] and Kamalâkara.[4] Jñânarâja simply observes:

"If the interpolator (of a Square-nature) be divided by the square of an optional number then its roots will be divided by that optional number."

Thus we have, in general, if $x = \alpha$, $y = \beta$ be a solution of the equation

$$Nx^2 \pm k = y^2,$$

[1] *BrSpSi*, xviii. 70.

[2] *BBi*, p. 34.

[3] *NBi*, I, R. 76-76½.

[4] *SiTVi*, xiii. 215.

$x = \alpha/m$, $y = \beta/m$ is a solution of the equation

$$Nx^2 \pm k/m^2 = y^2;$$

and $x = n\alpha$, $y = n\beta$ is a solution of the equation

$$Nx^2 \pm n^2k = y^2,$$

where m, n are arbitrary rational numbers.

By this Lemma, the solutions of the Square-natures

$$(i)\quad 6x^2 + 12 = y^2,$$
$$(ii)\quad 6x^2 + 75 = y^2,$$
$$\text{and}\quad (iii)\quad 6x^2 + 300 = y^2,$$

can be derived, as shown by Bhâskara II,[1] from those of

$$6x^2 + 3 = y^2,$$

since $12 = 2^2.3$, $75 = 5^2.3$, and $300 = 10^2.3$. How to solve this latter equation will be indicated later on.

Rational Solution. In order to obtain a first solution of $Nx^2 + 1 = y^2$ the Hindus generally suggest the following tentative method: Take an arbitrary small rational number α, such that its square multiplied by the *guṇaka* N and increased or diminished by a suitably chosen rational number k will be an exact square. In other words, we shall have to obtain empirically a relation of the form

$$N\alpha^2 \pm k = \beta^2,$$

where α, k, β are rational numbers. This relation will be hereafter referred to as the *Auxiliary Equation*. Then by Brahmagupta's Corollary, we get from it the relation

$$N(2\alpha\beta)^2 + k^2 = (\beta^2 + N\alpha^2)^2,$$

or

$$N\left(\frac{2\alpha\beta}{k}\right)^2 + 1 = \left(\frac{\beta^2 + N\alpha^2}{k}\right)^2.$$

[1] *BBi*, p. 41.

Hence, one rational solution of the equation $Nx^2 + 1 = y^2$ is given by

$$x = \frac{2\alpha\beta}{k}, \quad y = \frac{\beta^2 + N\alpha^2}{k}. \tag{A}$$

Srîpati's Rational Solution. Śrîpati (1039) has shown how a rational solution of the Square-nature can be obtained more easily and directly without the intervention of an auxiliary equation. He says:

"Unity is the lesser root. Its square multiplied by the *prakṛti* is increased or decreased by the *prakṛti* combined with an (optional) number whose square-root will be the greater root. From them will be obtained two roots by the Principle of Composition."[1]

If m^2 be a rational number optionally chosen, we have the identity

$$N.1^2 + (m^2 - N) = m^2,$$

or

$$N.1^2 - (N - m^2) = m^2.$$

Then, applying Brahmagupta's Corollary to either, we get

$$N(2m)^2 + (m^2 \sim N)^2 = (m^2 + N)^2;$$

$$\therefore \quad N\left(\frac{2m}{m^2 \sim N}\right)^2 + 1 = \left(\frac{m^2 + N}{m^2 \sim N}\right)^2.$$

Hence

$$x = \frac{2m}{m^2 \sim N}, \quad y = \frac{m^2 + N}{m^2 \sim N}, \tag{B}$$

where m is any rational number, is a solution of the equation

$$Nx^2 + 1 = y^2.$$

The above solution reappears in the works of later Hindu algebraists. Bhâskara II says:

[1] *SiŚe*, xiv. 33.

"Or divide twice an optional number by the difference between the square of that optional number and the *prakṛti*. This (quotient) will be the lesser root (of a Square-nature) when unity is the additive. From that (follows) the greater root."[1]

Nârâyaṇa states :

"Twice an optional number divided by the difference between the square of that optional number and the *guṇaka* will be the lesser root. From that with the additive unity determine the greater root."[2]

Similar statements are found also in the works of Jñânarâja and Kamalâkara.[3]

If m be an optional number, it is stated that $\frac{2m}{m^2 \sim N}$ is a lesser root of $Nx^2 + 1 = y^2$. Then, substituting that value of x in the equation, we get

$$y^2 = N\left(\frac{2m}{m^2 \sim N}\right)^2 + 1,$$

$$= \left(\frac{m^2 + N}{m^2 \sim N}\right)^2.$$

Hence the greater root is

$$y = \frac{m^2 + N}{m^2 \sim N}.$$

The same solution will be obtained by assuming

$$y = mx - 1.$$

Kṛṣṇa points out that it can also be found thus :

$$4Nm^2 = (m^2 + N)^2 - (m^2 \sim N)^2, \text{ identically.}$$

$$\therefore \quad 4Nm^2 + (m^2 \sim N)^2 = (m^2 + N)^2,$$

[1] *BBi*, p. 34. [2] *NBi*, I, R. 77f.

[3] *SiTVi*, xiii. 216.

or $$N\left(\frac{2m}{m^2 \sim N}\right)^2 + 1 = \left(\frac{m^2 + N}{m^2 \sim N}\right)^2.$$

His remark that this method does not require the help of the Principle of Composition shows that Bhâskara II and others obtained the solution in the way indicated by Srîpati.

The above rational solution of the Square-nature has been hitherto attributed by modern historians of mathematics to Bhâskara II. But it is now found to be due to an anterior writer, Srîpati (1039). It was rediscovered in Europe by Brouncker (1657).

Illustrative Examples. In illustration of the foregoing rules we give the following examples with their solutions from Bhâskara II.

Examples. "Tell me, O mathematician, what is that square which multiplied by 8 becomes, together with unity, a square; and what square multiplied by 11 and increased by unity, becomes a square."[1]

That is to say, we have to solve

$$(1) \quad 8x^2 + 1 = y^2,$$
$$(2) \quad 11x^2 + 1 = y^2.$$

Solutions. "In the second example assume 1 as the lesser root. Multiplying its square by the *prakṛti*, namely 11, subtracting 2 and then extracting the square-root, we get the greater root as 3. Hence the statement for composition is[2]

$$m = 11 \qquad l = 1 \qquad g = 3 \qquad i = -2$$
$$l = 1 \qquad g = 3 \qquad i = -2$$

[1] *BBi*, p. 35.

[2] The abbreviations are: m = multiplier, l = lesser root, g = greater root and i = interpolator. In the original they are respectively *pra*, *ka*, *jye*, and *kṣe*, the initial syllables of the corresponding Sanskrit terms.

Proceeding as before we obtain the roots for the additive 4: $l = 6$, $g = 20$, (for) $i = 4$. Then by the rule, 'If the interpolator (of a Square-nature) divided by the square of an optional number etc.,'[1] are found the roots for the additive unity: $l = 3$, $g = 10$ (for) $i = 1$. Whence by the Principle of Composition of Equals, we get the lesser and greater roots: $l = 60$, $g = 199$ (for) $i = 1$. In this way an infinite number of roots can be deduced.

"Or, assuming 1 for the lesser root, we get for the additive 5: $l = 1$, $g = 4$, (for) $i = 5$. Whence by the Principle of Composition of Equals, the roots are $l = 8$, $g = 27$, (for) $i = 25$. Then by the rule, 'If the interpolator (of a Square-nature) divided by the square of an optional number etc.,' taking 5 as the optional number, we get the roots for the additive unity: $l = 8/5$, $g = 27/5$, (for) $i = 1$. The statement of these for composition with the previous roots is

$$\begin{array}{llll} m = 11 & l = 8/5 & g = 27/5 & i = 1 \\ & l = 3 & g = 10 & i = 1 \end{array}$$

By the Principle of Composition the roots are obtained as: $l = 161/5$, $g = 534/5$ (for) $i = 1$.

"Or composing according to the rule, 'The difference of the two cross-products is a lesser root etc.,' we get the roots: $l = 1/5$, $g = 6/5$ (for) $i = 1$. And so on in many ways.

"The two roots for the additive unity will now be found in a different way by the rule, 'Or divide twice an optional number by the difference between the square of that optional number and the *prakṛti* etc.' Here, in the first example, assume the optional number to be 3. Its square is 9; multiplier is 8; their difference is 1;

[1] *Vide supra*, p. 151.

dividing by this twice the optional number, namely 6, we get the lesser root for the additive unity as 6. Whence, proceeding as before, the greater root comes out as 17.

"In the same way, in the second example also, assuming the optional number to be 3, the lesser and greater roots are found to be (3, 10).

"Thus, by virtue of (the infinite variety of) the optional values as well as of (the infinitely repeated application of) the Principle of Additive and Subtractive Compositions, an infinite number of roots (may be found)."[1]

Solution in Positive Integers. As has been stated before, the aim of the Hindus was to obtain solutions of the Square-nature in positive integers; so its first solution must be integral. But neither the tentative method of Brahmagupta nor that of Srîpati is of much help in this direction, for they do not *always* yield the desired result. These authors, however, discovered that if the interpolator of the auxiliary equation in the tentative method be ± 1, ± 2 or ± 4, an integral solution of the equation $Nx^2 + 1 = y^2$ can always be found. Thus Srîpati (1039) expressly observes, "If 1, 2 or 4 be the additive or subtractive (of the auxiliary equation) the lesser and greater roots will be integral (*abhinna*).[2]

(*i*) If $k = \pm 1$; then the auxiliary equation will be[3]

$$N\alpha^2 \pm 1 = \beta^2,$$

[1] The original is, "Evamiṣṭavaśât samâsântarabhâvanâbhyâṁ ca padânâmânantyaṃ." (*BBi*, p. 36).

[2] "Dvyekâmbudhikṣepaviśodhanâbhyâṃ
Syâtâmabhinne laghuvṛddhamûle."—*SiSe*, xiv. 32.

The Sanskrit word *abhinna* literally means "non-fractional."

[3] The special treatment of the equation $Nx^2 - 1 = y^2$ is given later on.

where α, β are integers. Then by Brahmagupta's Corollary, we get

$$x = 2\alpha\beta, \quad y = \beta^2 + N\alpha^2$$

as the required first solution in positive integers of the equation $Nx^2 + 1 = y^2$.

(*ii*) Let $k = \pm 2$; then the auxiliary equation is

$$N\alpha^2 \pm 2 = \beta^2.$$

By Brahmagupta's Corollary, we have

$$N(2\alpha\beta)^2 + 4 = (\beta^2 + N\alpha^2)^2,$$

or $$N(\alpha\beta)^2 + 1 = \left(\frac{\beta^2 + N\alpha^2}{2}\right)^2.$$

Hence the required first solution is

$$x = \alpha\beta, \quad y = \tfrac{1}{2}(\beta^2 + N\alpha^2).$$

Since $$N\alpha^2 = \beta^2 \mp 2,$$

we have $$\tfrac{1}{2}(\beta^2 + N\alpha^2) = \beta^2 \mp 1 = \text{a whole number}.$$

(*iii*) Now suppose $k = +4$; so that

$$N\alpha^2 + 4 = \beta^2.$$

With an auxiliary equation like this the first integral solution of the equation $Nx^2 + 1 = y^2$ is

$$x = \tfrac{1}{2}\alpha\beta,$$
$$y = \tfrac{1}{2}(\beta^2 - 2);$$

if α is even; or

$$x = \tfrac{1}{2}\alpha(\beta^2 - 1),$$
$$y = \tfrac{1}{2}\beta(\beta^2 - 3);$$

if β is odd.

Thus Brahmagupta says:

"In the case of 4 as additive the square of the second root diminished by 3, then halved and multiplied by the second root will be the (required) second root; The square of the second root diminished by unity and then divided by 2 and multiplied by the first root will be the (required) first root (for the additive unity)."[1]

The *rationale* of this solution is as follows:

Since $$N\alpha^2 + 4 = \beta^2, \tag{1}$$

we have $$N\left(\frac{\alpha}{2}\right)^2 + 1 = \left(\frac{\beta}{2}\right)^2. \tag{2}$$

Then, by Brahmagupta's Corollary, we get

$$N\left(\frac{\alpha\beta}{2}\right)^2 + 1 = \left(\frac{\beta^2}{4} + N\frac{\alpha^2}{4}\right)^2.$$

Substituting the value of N in the right-hand side expression from (1), we have

$$N\left(\frac{\alpha\beta}{2}\right)^2 + 1 = \left(\frac{\beta^2 - 2}{2}\right)^2. \tag{3}$$

Composing (2) and (3),

$$N\left\{\frac{\alpha}{2}(\beta^2 - 1)\right\}^2 + 1 = \left\{\frac{\beta}{2}(\beta^2 - 3)\right\}^2.$$

Hence $$x = \tfrac{1}{2}\alpha\beta, \quad y = \tfrac{1}{2}(\beta^2 - 2);$$

and $$x = \tfrac{1}{2}\alpha(\beta^2 - 1), \quad y = \tfrac{1}{2}\beta(\beta^2 - 3);$$

are solutions of

$$Nx^2 + 1 = y^2.$$

If β be even, the first values of (x, y) are integral. If β be odd, the second values are integral.

(*iv*) Finally, suppose $k = -4$; the auxiliary equation is

$$N\alpha^2 - 4 = \beta^2.$$

Then the required first solution in positive integers of $Nx^2 + 1 = y^2$ is

$$x = \tfrac{1}{2}\alpha\beta(\beta^2 + 3)(\beta^2 + 1),$$
$$y = (\beta^2 + 2)\{\tfrac{1}{2}(\beta^2 + 3)(\beta^2 + 1) - 1\}.$$

[1] *BrSpSi*, xviii. 67.

Brahmagupta says :

"In the case of 4 as subtractive, the square of the second is increased by three and by unity; half the product of these sums and that as diminished by unity (are obtained). The latter multiplied by the first sum less unity is the (required) second root; the former multiplied by the product of the (old) roots will be the first root corresponding to the (new) second root."[1]

The *rationale* of this solution is as follows :

$$N\alpha^2 - 4 = \beta^2. \qquad (1)$$

$$\therefore \quad N\left(\frac{\alpha}{2}\right)^2 - 1 = \left(\frac{\beta}{2}\right)^2.$$

Hence by Brahmagupta's Corollary, we get

$$N\left(\frac{\alpha\beta}{2}\right)^2 + 1 = \left(\frac{\beta^2}{4} + N\frac{\alpha^2}{4}\right)^2$$

$$= \{\tfrac{1}{2}(\beta^2 + 2)\}^2. \qquad (2)$$

Again, applying the Corollary, we have

$$N\{\tfrac{1}{2}\alpha\beta(\beta^2 + 2)\}^2 + 1 = \{\tfrac{1}{2}(\beta^4 + 4\beta^2 + 2)\}^2. \qquad (3)$$

Now, by the Lemma, we obtain from (2) and (3)

$$N\{\tfrac{1}{2}\alpha\beta(\beta^2 + 3)(\beta^2 + 1)\}^2 + 1$$

$$= [(\beta^2 + 2)\{\tfrac{1}{2}(\beta^2 + 3)(\beta^2 + 1) - 1\}]^2.$$

Hence $\quad x = \tfrac{1}{2}\alpha\beta(\beta^2 + 3)(\beta^2 + 1),$

$$y = (\beta^2 + 2)\{\tfrac{1}{2}(\beta^2 + 3)(\beta^2 + 1) - 1\},$$

is a solution of $Nx^2 + 1 = y^2$.

It can be proved easily that these values of x, y are integral. For, if β is even, $\beta^2 + 2$ is also even. Therefore, the above values of x, y are integral. If on the contrary β is odd, β^2 is also odd; then $\beta^2 + 1$ and $\beta^2 + 3$ are even. Hence in this case also the above values are integral.

[1] *BrSpSi*, xviii. 68.

Putting $p = \alpha\beta$, $q = \beta^2 + 2$ we can write the above solution in the form

$$x = \tfrac{1}{2}p(q^2 - 1),$$
$$y = \tfrac{1}{2}q(q^2 - 3),$$

in which it was found by Euler.

17. CYCLIC METHOD

Cyclic Method. It has been just shown that the most fundamental step in Brahmagupta's method for the general solution in positive integers of the equation

$$Nx^2 + 1 = y^2,$$

where N is a non-square integer, is to form an auxiliary equation of the kind

$$Na^2 + k = b^2,$$

where a, b are positive integers and $k = \pm 1, \pm 2$ or ± 4. For, from that auxiliary equation, by the Principle of Composition, applied repeatedly whenever necessary, one can derive, as shown above, one positive integral solution of the original Square-nature. And thence, again by means of the same principle, an infinite number of other solutions in integers can be obtained. How to form an auxiliary equation of this type was a problem which could not be solved completely and satisfactorily by Brahmagupta. In fact, he could not do it otherwise than by trial. But Bhâskara II succeeded in evolving a very simple and elegant method by means of which one can derive an auxiliary equation having the required interpolator $\pm 1, \pm 2$ or ± 4, simultaneously with its two integral roots, from another auxiliary equation empirically formed with any simple integral value of the interpolator, positive or negative. This method is called by the technical name *Cakravâla* or the "Cyclic Method."[1]

The purpose of the Cyclic Method has been defined by Bhâskara II thus: "By this method, there will appear two integral roots corresponding to an equation with ± 1, ± 2 or ± 4 as interpolator."[2]

Bhâskara's Lemma. The Cyclic Method of Bhâskara II is based upon the following Lemma:

If $$Na^2 + k = b^2,$$

where a, b, k are integers, k being positive or negative, then

$$N\left(\frac{am+b}{k}\right)^2 + \frac{m^2-N}{k} = \left(\frac{bm+Na}{k}\right)^2,$$

where m is an arbitrary whole number.

The *rationale* of this Lemma is simple: We have

$$Na^2 + k = b^2,$$

and $$N.1^2 + (m^2 - N) = m^2, \text{ identically.}$$

Then by Brahmagupta's Lemma, we get

$$N(am+b)^2 + k(m^2 - N) = (bm + Na)^2.$$

$$\therefore \quad N\left(\frac{am+b}{k}\right)^2 + \frac{m^2-N}{k} = \left(\frac{bm+Na}{k}\right)^2$$

Bhâskara's Rule. Bhâskara II (1150) says:

"Considering the lesser root, greater root and interpolator (of a Square-nature) as the dividend, addend and divisor (respectively of a pulveriser), the (indeterminate) multiplier of it should be so taken as will make the residue of the *prakṛti* diminished by the square of that multiplier or the latter minus the *prakṛti* (as the case

[1] The Sanskrit word *Çakravâla* means "circle," especially "horizon." The method is so called, observes Sûryadâsa, because it proceeds as in a circle, the same set of operations being applied again and again in a continuous round.

[2] *BBi*, p. 38.

may be) the least. That residue divided by the (original) interpolator is the interpolator (of a new Square-nature); it should be reversed in sign in case of subtraction from the *prakṛti*. The quotient corresponding to that value of the multiplier is the (new) lesser root; thence the greater root. The same process should be followed repeatedly putting aside (each time) the previous roots and the interpolator. This process is called *Cakravâla* (or the 'Cyclic Method').[1] By this method, there will appear two integral roots corresponding to an equation with ± 1, ± 2 or ± 4 as interpolator. In order to derive integral roots corresponding to an equation with the additive unity from those of the equation with the interpolator ± 2 or ± 4 the Principle of Composition (should be applied)."[2]

Suppose we have an equation of the form

$$Na^2 + k = b^2, \tag{1}$$

where a, b, k are simple integers, relatively prime, k being positive or negative. Then by Bhâskara's Lemma

$$N\left(\frac{am+b}{k}\right)^2 + \frac{m^2-N}{k} = \left(\frac{bm+Na}{k}\right)^2, \tag{2}$$

where m is an arbitrary integral number. In the above rule, m has been styled the indeterminate multiplier. Now, by means of the pulveriser, its value is determined so that

$$\frac{am+b}{k} \text{ is a whole number.}$$

[1] The original text is *cakravâlamidaṁ jaguḥ*. The commentator Kṛṣṇa explains, "âcâryâ etadgaṇitaṁ cakravâlamiti jaguḥ" or "The learned professors call this method of calculation the *Cakravâla*." So Bhâskara II appears to have taken the Cyclic Method from earlier writers. But it is not found in any work anterior to him so far known.

[2] *BBi*, pp. 36ff.

Again, of the various such values, Bhâskara II chooses that one which will make $|m^2 - N|$ as small as possible. Let that value of m be n. Now let

$$a_1 = \frac{an + b}{k},$$

$$b_1 = \frac{bn + Na}{k},$$

$$k_1 = \frac{n^2 - N}{k}.$$

The numbers a_1, b_1, k_1 are all integral. The equation (2) then becomes

$$Na_1^2 + k_1 = b_1^2. \qquad (3)$$

Proceeding exactly in the same way, we can obtain from (3) a new equation of the same kind,

$$Na_2^2 + k_2 = b_2^2,$$

where again a_2, b_2, k_2 are whole numbers. By repeating the process, we shall ultimately arrive at an equation, states Bhâskara II, in which the interpolator k will reach the value $\pm$ 1, $\pm$ 2 or $\pm$ 4, and in which (a, b) will be integers.

Nârâyaṇa's Rule. The above rule of Bhâskara II has been reproduced by Nârâyaṇa (1350). He writes:

"Making the lesser root, greater root and interpolator (of a Square-nature) the dividend, addend and divisor (respectively of a pulveriser), the (indeterminate) multiplier of it should be determined in the way described before. The *prakṛti* being subtracted from the square of that or the square of the multiplier being subtracted from the *prakṛti*, the remainder divided by the (original) interpolator is the interpolator (of a new Square-nature); and it will be reversed in sign in case of subtraction of the square of the multiplier. The quotient (corresponding to that value of the multiplier) is the lesser root (of the new Square-nature); and that multiplied by the multiplier and diminished by the product of the previous lesser root and (new) interpolator will be its greater root. By doing so repeatedly will be obtained two integral roots corresponding to the interpolator $\pm 1, \pm 2$ or ± 4. In order to derive integral roots for the additive unity from those answering to the interpolator ± 2 or ± 4, the Principle of Composition (should be adopted)."[1]

It will be noticed that Nârâyaṇa does not expressly state that the value of the indeterminate multiplier m should be so chosen as will make $|m^2 - N|$ least. It is perhaps particularly noteworthy that he recognised the relation

$$b_1 = a_1 n - k_1 a.$$

For

$$b_1 = \frac{bn + Na}{k},$$

$$= \frac{n(a_1 k - an) + Na}{k}, \quad [\because \quad a_1 k = an + b]$$

$$= a_1 n - \left(\frac{n^2 - N}{k}\right) a,$$

$$= a_1 n - k_1 a,$$

$$\therefore \quad a = \frac{a_1 n - b_1}{k_1}.$$

Similarly, it will be found that

$$b = \frac{b_1 n - N a_1}{k_1}.$$

For

$$b_1 n = a_1 n^2 - k_1 a n,$$

[1] *NBi*, I, R. 79-82.

$$= a_1(N + kk_1) - k_1an, \qquad [\because\ kk_1 = n^2 - N]$$

$$= a_1N + k_1b, \qquad [\because\ a_1k = an + b]$$

$$\therefore \quad b = \frac{b_1n - Na_1}{k_1}.$$

Illustrative Examples. In illustration of the Cyclic Method, Bhâskara II works out in detail the following examples :

"What is that number whose square multiplied by 67 or 61 and then added by unity becomes capable of yielding a square-root ? Tell me, O friend, if you have a thorough knowledge of the method of the Square-nature."[1]

That is to say, we are to solve

$$(i) \quad 67x^2 + 1 = y^2,$$

$$(ii) \quad 61x^2 + 1 = y^2.$$

Leaving out the details of the operations in connection with the process of the pulveriser, Bhâskara's solutions are substantially as follows :

$$(i) \quad 67x^2 + 1 = y^2.$$

We take the auxiliary equation

$$67 . 1^2 - 3 = 8^2.$$

Then, by the Lemma,

$$67\left(\frac{1 . m + 8}{-3}\right) + \frac{m^2 - 67}{-3} = \left(\frac{8m + 67 . 1}{-3}\right)^2. \quad (1)$$

By the method of the *Kuṭṭaka* the solution of

$$\frac{m + 8}{-3} = \text{an integer},$$

[1] *BBi*, p. 38.

It is remarkable that the equation $61x^2 + 1 = y^2$ was proposed by Fermat to Frénicle in a letter of February, 1657. Euler solved it in 1732.

is $m = -3t + 1$. Putting $t = -2$, we get $m = 7$ which makes $|m^2 - 67|$ least. On substituting this value, the equation (1) reduces to

$$67 . 5^2 + 6 = 41^2.$$

Again, by the Lemma, we have

$$67\left(\frac{5n+41}{6}\right)^2 + \frac{n^2-67}{6} = \left(\frac{41n+67.5}{6}\right)^2. \quad (2)$$

The solution of

$$\frac{5n+41}{6} = \text{a whole number,}$$

is $n = 6t + 5$. $|n^2 - 67|$ will be least for the value $t = 0$, that is, when $n = 5$. The equation (2) then becomes

$$67 . 11^2 - 7 = 90^2.$$

Now, we form

$$67\left(\frac{11p+90}{-7}\right)^2 + \frac{p^2-67}{-7} = \left(\frac{90p+67.11}{-7}\right)^2. \quad (3)$$

The solution of

$$\frac{11p+90}{-7} = \text{an integral number,}$$

is $p = -7t + 2$. Taking $t = -1$, we have $p = 9$; and this value makes $|p^2 - 67|$ least. Substituting that in (3) we get

$$67 . 27^2 - 2 = 221^2.$$

By the Principle of Composition of Equals, we get from this equation

$$67\,(2 . 27 . 221)^2 + 4 = (221^2 + 67 . 27^2)^2,$$

or $$67(11934)^2 + 4 = (97684)^2.$$

Dividing out by 4, we have

$$67\,(5967)^2 + 1 = (48842)^2.$$

Hence $x = 5967$, $y = 48842$ is a solution of (*i*).

(*ii*) $\quad 61x^2 + 1 = y^2.$

Here we start with the auxiliary equation

$$61 \cdot 1^2 + 3 = 8^2.$$

By the Lemma, we have

$$61\left(\frac{m+8}{3}\right)^2 + \frac{m^2 - 61}{3} = \left(\frac{8m + 61}{3}\right)^2. \qquad (1)$$

Now the solution of

$$\frac{m+8}{3} = \text{an integer},$$

is $m = 3t + 1$. Putting $t = 2$, we get the value $m = 7$ which makes $|m^2 - 61|$ least. On substituting this value in (1), it becomes

$$61 \cdot 5^2 - 4 = 39^2.$$

Dividing out by 4, we get

$$61(\tfrac{5}{2})^2 - 1 = (\tfrac{39}{2})^2. \qquad (2)$$

By the Principle of Composition of Equals, we have

$$61(2 \cdot \tfrac{5}{2} \cdot \tfrac{39}{2})^2 + 1 = \{(\tfrac{39}{2})^2 + 61(\tfrac{5}{2})^2\}^2,$$

or

$$61(\tfrac{195}{2})^2 + 1 = (\tfrac{1523}{2})^2. \qquad (3)$$

Combining (2) and (3),

$$61(3805)^2 - 1 = (29718)^2.$$

Composing this with itself, we get

$$61(226153980)^2 + 1 = (1766319049)^2.$$

Hence $x = 226153980$, $y = 1766319049$ is a solution of (*ii*).

The following two examples have been cited by Nârâyaṇa :

(*iii*) $\quad 103x^2 + 1 = y^2,$

(*iv*) $\quad 97x^2 + 1 = y^2.$

Their solutions are given substantially as follows:

For (*iii*) we have the auxiliary equation

$$103 \cdot 1^2 - 3 = 10^2.$$

By the Lemma, we get

$$103\left(\frac{m+10}{-3}\right)^2 + \frac{m^2 - 103}{-3} = \left(\frac{10m + 103}{-3}\right)^2.$$

The general solution of

$$\frac{m+10}{-3} = \text{an integer},$$

is $m = -3t + 2$. Putting $t = -3$, we get $m = 11$. Then

$$103 \cdot 7^2 - 6 = 71^2.$$

Again, by the Lemma,

$$103\left(\frac{7n+71}{-6}\right)^2 + \frac{n^2 - 103}{-6} = \left(\frac{71n + 103 \cdot 7}{-6}\right)^2.$$

The solution of

$$\frac{7n+71}{-6} = \text{a whole number},$$

is $n = -6t + 1$. Taking $t = -1$, we get

$$103 \cdot 20^2 + 9 = 203^2.$$

Next, we have

$$103\left(\frac{20p + 203}{9}\right)^2 + \frac{p^2 - 103}{9} = \left(\frac{203p + 103 \cdot 20}{9}\right)^2.$$

Now, $$\frac{20p + 203}{9} = \text{an integral number}$$

for $p = 9t + 2$. When $t = 1$, $p = 11$. On taking this value we find

$$103 \cdot 47^2 + 2 = 477^2.$$

Applying the Principle of Composition of Equals, we get

$$103(2.47.477)^2 + 4 = (477^2 + 103.47^2)^2,$$

or $$103(44838)^2 + 4 = (455056)^2.$$

Hence $$103(22419)^2 + 1 = (227528)^2,$$

which gives $x = 22419$, $y = 227528$ as a solution of (*iii*).

For the solution of (*iv*) the auxiliary equation is

$$97 \cdot 1^2 + 3 = 10^2.$$

Therefore

$$97\left(\frac{m + 10}{3}\right)^2 + \frac{m^2 - 97}{3} = \left(\frac{10m + 97}{3}\right)^2.$$

The solution of

$$\frac{m + 10}{3} = \text{an integer},$$

is $m = 3t + 2$. Taking $t = 3$, we have $m = 11$. Then

$$97 \cdot 7^2 + 8 = 69.^2$$

Next, we have

$$97\left(\frac{7n + 69}{8}\right)^2 + \frac{n^2 - 97}{8} = \left(\frac{69n + 97 \cdot 7}{8}\right)^2.$$

The solution of

$$\frac{7n + 69}{8} = \text{an integer},$$

is $n = 8t + 5$. Taking $t = 1$, that is, $n = 13$, we get

$$97 \cdot 20^2 + 9 = 197^2.$$

Whence

$$97\left(\frac{20p + 197}{9}\right)^2 + \frac{p^2 - 97}{9} = \left(\frac{197p + 97 \cdot 20}{9}\right)^2.$$

The solution of

$$\frac{20p + 197}{9} = \text{a whole number},$$

is $p = 9t + 5$. Putting $t = 1$, we get $p = 14$. With this value of p we have

$$97 \cdot 53^2 + 11 = 522^2.$$

Whence

$$97\left(\frac{53q + 522}{11}\right)^2 + \frac{q^2 - 97}{11} = \left(\frac{522q + 97 \cdot 53}{11}\right)^2.$$

The solution of

$$\frac{53q + 522}{11} = \text{an integer},$$

is $q = 11t + 8$. The appropriate value of q is given by $t = 0$. So, taking $q = 8$, we have

$$97 \cdot 86^2 - 3 = 847^2.$$

Next, we find

$$97\left(\frac{86r + 847}{-3}\right) + \frac{r^2 - 97}{-3} = \left(\frac{847r + 97 \cdot 86}{-3}\right)^2.$$

The solution of

$$\frac{86r + 847}{-3} = \text{a whole number},$$

is $r = 3t + 1$. Putting $t = -3$, we get $r = 10$. Taking this value, we have

$$97 \cdot 569^2 - 1 = 5604^2.$$

By the Principle of Composition of Equals, we find

$$97(6377352)^2 + 1 = (62809633)^2.$$

Hence $x = 6377352$, $y = 62809633$ is a solution of (*iv*).

Proofs. It has been stated by Bhâskara II that:

(1) when a_1 is an integer, k_1 and b_1 are each a whole number;

(2) his Cyclic Method will in every case lead to the desired result.

He has not adduced proofs. We presume that he

knew a proof at least of the first proposition. For he must have recognised the simple relation

$$b_1 = a_1 n - k_1 a,$$

which has been expressly stated by Nârâyaṇa (1350). This shows at once that b_1 will be a whole number, if k_1 is so. This is also evident from the equation, $Na_1^2 + k_1 = b_1^2$, itself. Hence, it now remains to prove that k_1 is an integral number.

Eliminating b between

$$a_1 = \frac{an + b}{k},$$

and

$$b_1 = \frac{bn + Na}{k},$$

we have

$$k(a_1 n - b_1) = a(n^2 - N),$$

or

$$\frac{k}{a}(a_1 n - b_1) = n^2 - N.$$

Therefore

$$\frac{k}{a}(a_1 n - b_1) \text{ is an integer.}$$

Since k and a have no common factor, a must divide $a_1 n - b_1$; that is

$$\frac{a_1 n - b_1}{a} = \frac{n^2 - N}{k} = k_1 = \text{an integer}.$$

Hence b_1 also is a whole number.[1]

[1] *Hankel's Proof*: Hankel proves these two results thus:

Since $$a_1 k = an + b \text{ and } k = b^2 - Na^2,$$

we get $$a_1(b^2 - Na^2) = an + b,$$

or $$\frac{b}{a}(a_1 b - 1) = (n + Naa_1).$$

Since a, b have no common factor, a must divide $a_1 b - 1$; that is,

$$\frac{a_1 b - 1}{a} = \text{an integer}.$$

18. SOLUTION OF $Nx^2 \pm c = y^2$

The general solution of the indeterminate quadratic equation

$$Nx^2 \pm c = y^2$$

in positive integers was first given by Brahmagupta (628). He says:

"From two roots (of a Square-nature) with any given additive or subtractive, by making (combination) with the roots for the additive unity, other first and second roots (of the equation having) the given additive or subtractive (can be found)."[1]

Eliminating n between

$$a_1 k = an + b, \quad b_1 k = bn + Na,$$

we get $$a_1 b - ab_1 = 1.$$

Hence $$b_1 = \frac{a_1 b - 1}{a} = \text{a whole number}.$$

Now $$n^2 - N = \frac{(a_1 k - b)^2 - Na^2}{a^2}$$

$$= \frac{a_1^2 k^2 - 2bka_1 + k}{a^2}$$

$$= \frac{k(a_1^2 k - 2ba_1 + 1)}{a^2}.$$

Therefore $$\frac{k}{a^2}(a_1^2 k - 2ba_1 + 1) \text{ is a whole number.}$$

Since a, k have no common factor, it follows that

$$\frac{a_1^2 k - 2ba_1 + 1}{a^2} = \frac{n^2 - N}{k} = k_1 = \text{an integer.}$$

Also $$k_1 = \frac{n^2 - N}{k} = \frac{a_1^2 k - 2ba_1 + 1}{a^2}$$

$$= \frac{a_1^2(b^2 - Na^2) - 2ba_1 + 1}{a^2}$$

$$= \left(\frac{a_1 b - 1}{a}\right)^2 - Na_1^2.$$

[1] *BrSpSi*, xviii, 66.

Thus having known a single solution in positive integers of the equation $Nx^2 \pm c = y^2$, says Brahmagupta, an infinite number of other integral solutions can be obtained by making use of the integral solutions of $Nx^2 + 1 = y^2$. If (p, q) be a solution of the former equation found empirically and if (α, β) be an integral solution of the latter then, by the Principle of Composition,

$$x = p\beta \pm q\alpha, \quad y = q\beta \pm Np\alpha$$

will be a solution of the former. Repeating the operations we can easily deduce as many solutions as we like.

This method reappears in later Hindu algebras. Bhâskara II says:

"In (a Square-nature) with the additive or subtractive greater (than unity), one should find two roots by his own intelligence only; then by their composition with the roots obtained for the additive unity an infinite number of roots (will be found)."[1]

Nârâyaṇa writes similarly:

"When the additive or subtractive is greater than unity, two roots should be determined by one's own intelligence. Then, by combining them with the roots for the additive unity, an infinite number of roots can be obtained."[2]

We take the following illustrative examples with solutions from Nârâyaṇa:

Example. "Tell me that square which being multiplied by 13 and then increased or diminished by 17 or 8 becomes capable of yielding a square root."[3]

[1] *BBi*, p. 42.

[2] *NBi*, I, R. 86.

[3] *NBi*, I, *Ex*. 44.

That is, solve

$$(1)\quad 13x^2 \pm 17 = y^2,$$
$$(2)\quad 13x^2 \pm 8 = y^2.$$

Solution. "In the first example it is stated that the multiplier = 13 and interpolator = 17.

"Now the roots for the interpolator 3 are (1, 4). And for the interpolator 51, the roots are (1, 8). For the composition of these with the previous roots (1, 4) the statement will be

$$m = 13 \quad l = 1 \quad g = 8 \quad i = 51$$
$$l = 1 \quad g = 4 \quad i = 3$$

So, by the Addition Lemma, we get the roots corresponding to the interpolator 153 as (12, 45). The rule says, 'If the interpolator (of a Square-nature) be divided by the square of an optional number etc.' Now take the optional number to be 3, so that the interpolator may be reduced to 17. For $3^2 = 9$ and $153/9 = 17$. Therefore, dividing the roots just obtained by the optional number 3, we get the required roots (4, 15).

"Applying the Subtraction Lemma and proceeding similarly we get the roots for the interpolator 17 as (4/3, 19/3).

"In the second example the statement is : multiplier = 13, interpolator = − 17. Proceeding as before we get (by the Addition Lemma) the roots (147, 530); and (by the Subtraction Lemma), the roots (3, 10)."[1]

Form $Mn^2x^2 \pm c = y^2$. Brahmagupta says :

"If the multiplier is that divided by a square, the first root is that divided by its root."[2]

[1] Our MS. does not contain the solution of the equations $13x^2 \pm 8 = y^2$.

[2] *BrSpSi*, xviii. 70.

That is to say, suppose the equation to be

$$Mn^2x^2 \pm c = y^2, \tag{1}$$

so that the multiplier (*i.e.*, coefficient of x^2) is divisible by n^2. Putting $nx = u$, we get

$$Mu^2 \pm c = y^2. \tag{2}$$

Then clearly the first root of (1) is equal to the first root of (2) divided by n. The corresponding second root will be the same for both the equations.

The same rule is taught by Bhâskara II[1] and Nârâyaṇa. The latter says:

"Divide the multiplier (of a Square-nature) by an arbitrary square number so that there is left no remainder. Take the quotient as the multiplier (of another Square-nature). The lesser root (of the reduced equation) divided by the square-root of the divisor will be the lesser root (of the original equation)."[2]

Form $a^2x^2 \pm c = y^2$. For the solution of a Square-nature of this particular form, Brahmagupta gives the following rule:

"If the multiplier be a square, the interpolator divided by an optional number and then increased and decreased by it, is halved. The former (of these results) is the second root; and the other divided by the square-root of the multiplier is the first root."[3]

Thus, it is stated that

$$x = \frac{1}{2a}\left(\frac{\pm c}{m} - m\right),$$

$$y = \tfrac{1}{2}\left(\frac{\pm c}{m} + m\right),$$

[1] *BBi*, p. 42.
[2] *NBi*, I, R. 84.
[3] *BrSpSi*, xviii. 69.

where m is an arbitrary number, is a solution of the equation

$$a^2x^2 \pm c = y^2.$$

The same solution has been given by Bhâskara II and Nârâyaṇa.[1] Bhâskara's rule runs as follows:

"The interpolator divided by an optional number is set down at two places; the quotient is diminished (at one place) and increased (at the other) by that optional number and then halved. The former is again divided by the square-root of the multiplier. (The quotients) are respectively the lesser and greater roots."[2]

The *rationale* of the above solution has been given by the commentators Sûryadâsa and Kṛṣṇa substantially as follows:

$$\pm c = y^2 - a^2x^2$$
$$= (y - ax)(y + ax).$$

Assume $y - ax = m$, m being an arbitrary rational number. Then

$$y + ax = \frac{\pm c}{m}.$$

Whence by the rule of concurrence, we get

$$x = \frac{1}{2a}\left(\frac{\pm c}{m} - m\right),$$

$$y = \tfrac{1}{2}\left(\frac{\pm c}{m} + m\right).$$

Form $c - Nx^2 = y^2$. Though the equation of the form $c - Nx^2 = y^2$ has not been considered by any Hindu algebraist as deserving of special treatment, it occurs incidentally in examples. For instance, Bhâskara II has proposed the following problem:

[1] *NBi*, I, R. 85.

[2] *BBi*, p. 42.

"What is that square which being multiplied by -5 becomes, together with 21, a square? Tell me, if you know, the method (of solving the Square-nature) when the multiplier is negative."[1]

Thus it is required to solve

$$-5x^2 + 21 = y^2. \qquad (1)$$

Nârâyaṇa has a similar example, *viz.*,[2]

$$-11x^2 + 60 = y^2. \qquad (2)$$

Two obvious solutions of (1) are (1, 4) and (2, 1). Composing them with the roots of

$$-5x^2 + 1 = y^2,$$

says Bhâskara II, an infinite number of roots of (1) can be derived.

Form $Nx^2 - k^2 = y^2$. Bhâskara II observes:

"When unity is the subtractive the solution of the problem is impossible unless the multiplier is the sum of two squares."[3]

Nârâyaṇa writes:

"In the case of unity as the subtractive, the multiplier must be the sum of two squares. Otherwise, the solution is impossible."[4]

Thus it has been said that a rational solution of

$$Nx^2 - 1 = y^2,$$

and consequently of

$$Nx^2 - k^2 = y^2$$

is not possible unless N is the sum of two squares.

[1] *BBi*, p. 43. [2] *NBi*, I, *Ex.* 43.

[3] "Rûpaśuddhau khiloddiṣṭaṁ vargayogo guṇo na cet"—*BBi*, p. 40.

[4] *NBi*, I, R. 83.

For, if $x = p/q$, $y = r/s$ be a possible solution of the equation, we have

$$N(p/q)^2 - k^2 = (r/s)^2,$$

or $$N = (qr/ps)^2 + (qk/p)^2.$$

Bhâskara II then goes on :

"In case (the solution is) not impossible when unity is the subtractive, divide unity by the roots of the two squares and set down (the quotients) at two places. They are two lesser roots. Then find the corresponding greater roots at the two places. Or, when unity is the subtractive, the roots should be found as before."

Thus, according to Bhâskara II, two rational solutions of

$$Nx^2 - 1 = y^2,$$

where $N = m^2 + n^2$, will be

$$\left.\begin{array}{l} x = \dfrac{1}{m} \\ y = \dfrac{n}{m} \end{array}\right\}, \quad \left.\begin{array}{l} x = \dfrac{1}{n} \\ y = \dfrac{m}{n} \end{array}\right\}.$$

So two rational solutions of

$$(m^2 + n^2)\, x^2 - k^2 = y^2,$$

will be

$$\left.\begin{array}{l} x = \dfrac{k}{m} \\ y = \dfrac{kn}{m} \end{array}\right\}, \quad \left.\begin{array}{l} x = \dfrac{k}{n} \\ y = \dfrac{km}{n} \end{array}\right\}.$$

The following illustrative example of Bhâskara II[1] is also reproduced by Nârâyaṇa :[2]

$$13x^2 - 1 = y^2.$$

[1] *BBi*, p. 41.

[2] *NBi*, I, *Ex*. 58.

The former solves it substantially in the following ways:

(1) Since $13 = 2^2 + 3^2$ two rational solutions are $(1/2, 3/2)$ and $(1/3, 2/3)$.

(2) An obvious solution of

$$13x^2 - 4 = y^2$$

is $x = 1$, $y = 3$. Then dividing out by 4, as shown before, we get a solution of the equation $13x^2 - 1 = y^2$ as $(1/2, 3/2)$.

(3) Again, since an obvious solution of

$$13x^2 - 9 = y^2$$

is $x = 1, y = 2$, we get, on dividing out by 9, a solution of our equation as $(1/3, 2/3)$.

(4) From these fractional roots, we may derive *integral roots* by the Cyclic Method. Since

$$13(\tfrac{1}{2})^2 - 1 = (\tfrac{3}{2})^2,$$

we have, by Bhâskara's Lemma, m being an indeterminate multiplier,

$$13\left(\frac{m/2 + 3/2}{-1}\right)^2 + \frac{m^2 - 13}{-1} = \left(\frac{3m/2 + 13/2}{-1}\right)^2,$$

or
$$13\left(\frac{m + 3}{-2}\right)^2 + \frac{m^2 - 13}{-1} = \left(\frac{3m + 13}{-2}\right)^2.$$

The suitable value of m which will make $(m+3)/2$ an integer and $|m^2 - 13|$ minimum is 3. So that we have

$$13 \cdot 3^2 + 4 = 11^2.$$

From this again we get the relation

$$13\left(\frac{3n + 11}{4}\right)^2 + \frac{n^2 - 13}{4} = \left(\frac{11n + 13 \cdot 3}{4}\right)^2.$$

The appropriate value of the indeterminate multiplier in this case is $n = 3$. Substituting this value, we have

$$13.5^2 - 1 = 18^2.$$

Hence an integral solution of our equation $13x^2 - 1 = y^2$ is (5, 18).

"In all cases like this an infinite number of roots can be derived by composition with the roots for the additive unity."[1]

Nârâyaṇa states the methods (2) and (3) only.

19. GENERAL INDETERMINATE EQUATIONS OF THE SECOND DEGREE: SINGLE EQUATIONS

The earliest mention of the solution of the general indeterminate equation of the second degree is found in the *Bîjagaṇita* of Bhâskara II (1150). But there are good grounds to believe that he was not its first discoverer, for he is found to have taken from certain ancient authors a few illustrative examples the solutions of which presuppose a knowledge of the solution of such equations.[2] Neither those illustrations nor a treatment of equations of those types occurs in the algebra of Brahmagupta or in any other extant work anterior to Bhâskara II.

Bhâskara II distinguishes two kinds of indeterminate equations: *Sakṛt samîkaraṇa* (Single Equations) and *Asakṛt samîkaraṇa* (Multiple Equations).[3]

Solution. For the solution of the general indeterminate equation of the second degree, Bhâskara II (1150) lays down the following rule:

[1] "Iha sarvatra padânâṁ rûpakṣepapadâbhyâṁ bhâvanayâ'-nantyam"—*BBi*, p. 41.

[2] *Vide infra*, pp. 267f.

[3] *BBi*, pp. 106, 110.

"When the square, etc., of the unknown are present (in an equation), after the equi-clearance has been made, (find) the square-root of one side by the method described before for it, and the root of the other side by the method of the Square-nature. Then (apply) the method of (simple) equations to these roots. If (the other side) does not become a case for the Square-nature, then, putting it equal to the square of another unknown, the other side and so the value of the other (*i.e.*, the new) unknown should be obtained in the same way as in the Square-nature; and similarly the value of the first unknown. The intelligent should devise various artifices so that it may become a matter for (the application of) the Square-nature."[1]

He has further elucidated the rule thus:

"When, after the clearance of the two sides has been made, there remain the square, etc., of the unknown, then, by multiplying the two sides with a suitable number and by the help of other necessary operations as described before, the square-root of one side should be extracted. If there be present on the other side the square of the unknown with an absolute term, then the two roots of that side should be found by the method of the Square-nature. There the number associated with the square of the unknown is the *prakṛti* ('multiplier'), and the absolute number is to be considered as the interpolator. What is obtained as the lesser root in this way will be the value of the unknown associated with the multiplier (*prakṛti*); the greater root is (again) the root of that square (formed on the first side). Hence making an equation of this with the square-root of the first side, the value of the unknown on the first side should be determined.

[1] *BBi*, p. 99.

"But if there be present on the second side the square of the unknown together with (the first power of) the unknown, or only the (simple) unknown with or without an absolute number, then it is not a case for the Square-nature. How then is the root to be found in that case? So it has been said: 'If (the other side) does not become a case for the Square-nature etc.' Then, putting it equal to the square of another unknown, the square-root of one side should be found in the way indicated before, and the two roots of the other side should then be determined by the method of the Square-nature. There again the lesser root is the value of the unknown associated with the *prakṛti* and the greater root is equal to the square-root of that side of the equation. Forming proper equations with the roots, the values of the unknowns should be determined.

"If, however, even after the second side has been so treated, it does not turn out to be a case for the Square-nature, then the intelligent (mathematicians) should devise by their own sagacity all such artifices as will make it a case for the method of the Square-nature and then determine the values of the unknowns."[1]

Having thus indicated in a general way the broad outlines of his method for the solution of the general indeterminate equation of the second degree, Bhâskara II discusses the different types of equations severally, explaining the rules in every case in greater detail with the help of illustrative examples.

(*i*) *Solution of* $ax^2 + bx + c = y^2$

For the general solution of the quadratic indeterminate equation

$$ax^2 + bx + c = y^2, \qquad (1)$$

[1] *BBi*, p. 100.

Bhâskara II gives the following particular rule:

"On taking the square-root of one side, if there be on the second side only the square of the unknown together with an absolute number, in such cases, the greater and lesser roots should be determined by the method of the Square-nature. Of these two, the greater root is to be put equal to the square-root of the first side mentioned before, and thence the value of the first unknown should be determined. The lesser will be the value of the unknown associated with the *prakṛti*. In this way, the method of the Square-nature should be applied to this case by the intelligent."[1]

As an illustration of this rule Bhâskara II works out in detail the following example:

"What number being doubled and added to six times its square, becomes capable of yielding a square-root? O ye algebraist, tell it quickly."[2]

Solution. "Here let the number be x. Doubled and together with six times its square, it becomes $6x^2 + 2x$. This is a square. On forming an equation with the square of y, the statement is

$$6x^2 + 2x + 0y^2 = 0x^2 + 0x + y^2.$$

On making equi-clearance in this the two sides are $6x^2 + 2x$ and y^2.

"Then multiplying these two sides by 6 and superadding 1, the root of the first side, as described before, is $6x + 1$.

"Now on the second side of the equation remains $6y^2 + 1$. By the method of the Square-nature, its roots are: the lesser 2 and the greater 5, or the lesser 20 and the greater 49. Equating the greater root with the square-root of the first side, *viz.*, $6x + 1$, the value of

[1] *BBi*, pp. 100-1. [2] *BBi*, p. 101.

x is found to be 2/3 or 8. The lesser root, 2 or 20, is the value of y, the unknown associated with the *prakṛti*. In this way, by virtue of (the multiplicity of) the lesser and greater roots, many solutions can be obtained."[1]

In other words the method described above is this :

Completing the square on the left-hand side of the equation $ax^2 + bx + c = y^2$, we have

$$(ax + \tfrac{1}{2}b)^2 = ay^2 + \tfrac{1}{4}(b^2 - 4ac).$$

Putting $z = ax + \frac{1}{2}b$, $k = \frac{1}{4}(b^2 - 4ac)$, we get

$$ay^2 + k = z^2. \qquad (1.1)$$

If $y = l$, $z = m$ be found empirically to be a solution of this equation, another solution of it will be

$$y = lq \pm mp,$$
$$z = mq \pm alp,$$

where $ap^2 + 1 = q^2$. Hence a solution of (1) is

$$x = -\frac{b}{2a} + \frac{1}{a}(mq \pm alp),$$
$$y = lq \pm mp.$$

Now suppose $x = r$, when $z = m$; that is, let $m = ar + b/2$. Substituting in the above expressions, we get the required solution of (1) as

$$\left.\begin{aligned} x &= \frac{1}{2a}(bq - b) + qr \pm lp, \\ y &= lq \pm (apr + \tfrac{1}{2}bp); \end{aligned}\right\} \qquad (1.2)$$

where $ap^2 + 1 = q^2$ and $ar^2 + br + c = l^2$.

Thus having known *one* solution of $ax^2 + bx + c = y^2$, an infinite number of other solutions can be

[1] *BBi*, p. 101.

easily obtained by the method of Bhâskara II. The method is, indeed, a very simple and elegant one. It has been adopted by later Hindu algebraists. As the relevant portion of the algebra of Nârâyaṇa (1350) is now lost, we cannot reproduce his description of the method. Jñânarâja (1503) says :

"(Find) the square-root of the first side according to the method described before and, by the method of the Square-nature, the roots of the other side, where the coefficient of the square of the unknown is considered to be the *prakṛti* and the interpolator is an absolute term. Then the greater root will be equal to the previous square-root and the other (*i.e.*, the lesser root) to the unknown associated with the *prakṛti*."

The above solution (1.2), but with the upper sign only, was rediscovered in 1733 by Eluer.[1] His method is indirect and cumbrous. Lagrange's (1767) method begins in the same way as that of Bhâskara II. by completing the square on the left-hand side of the equation.[2]

(*ii*) *Solution of* $ax^2 + bx + c = a'y^2 + b'y + c'$

Bhâskara II has treated the more general type of quadratic indeterminate equations :

$$ax^2 + bx + c = a'y^2 + b'y + c'. \qquad (2)$$

His rule in this connection runs as follows :

"If there be the square of the unknown together with the (simple) unknown and an absolute number, putting it equal to the square of another unknown its root (should be investigated). Then on the other side (find)

[1] Leonard Euler, *Opera Mathematica*, vol. II, 1915, pp. 6-17; Compare also pp. 576-611.

[2] Additions to *Elements of Algebra by Leonard Euler*, translated into English by John Hewlett, 5th edition, London, 1840, pp. 537ff.

the roots by the method of the Square-nature, as has been stated before. Put the lesser root[1] equal to the root of the first side and the greater root equal to that of the second."[2]

He further elucidates the rule thus:

"In this case, on taking the square-root of the first side, there remain on the other side the square of the unknown and the (simple) unknown with or without an absolute number. In that case forming an equation of the second side with the square of another unknown, the roots (should be found). Of these (roots just determined), making the lesser equal to the root of the first side (of the given equation) and the greater to the root of the second side, the values of the unknowns should be determined."

Example. "Say what is the number of terms of a series (in A. P.) whose first term is 3, the common difference is 2; but whose sum multiplied by 3 is equal to the sum of a different number of terms."[3]

Solution. "Here the statements of the series are: first term = 3, common difference = 2, number of terms = x; first term = 3, common difference = 2, number of terms = y. The two sums are (respectively) $x^2 + 2x$, $y^2 + 2y$. Making three times the first equal to the second, the statement for clearance is

$$3x^2 + 6x = y^2 + 2y.$$

After the clearance, multiplying the two sides (of the equation) by 3 and superadding 9, the square-root of the first side is $3x + 3$. On the second side of the

[1] The meaning of the terms 'lesser root', 'greater root', etc., as used here, will be clear from the illustration and the general solution given below.

[2] *BBi*, p. 104. [3] *BBi*, p. 104.

equation stands $3y^2 + 6y + 9$. Forming an equation of this with z^2, and similarly multiplying the sides by 3 and superadding -18, the root of it is $3y + 3$. Then the roots of the other side, $3z^2 - 18$, by the method of the Square-nature are the lesser $= 9$ and greater $= 15$, or the lesser $= 13$ and greater $= 57$. Equating the lesser root with the square-root of the first side, namely, $3x + 3$, and the greater root with the square-root of the second side, namely, $3y + 3$, the values of x, y are found to be (2, 4) or (10, 18). So in every case."

In general, on completing the square on the left-hand side, equation (2) becomes

$$(ax + \tfrac{1}{2}b)^2 = aa'y^2 + ab'y + ac' + (\tfrac{1}{4}b^2 - ac).$$

Put
$$ax + \tfrac{1}{2}b = z, \tag{2.1}$$

and then complete the square on the right-hand side. Thus the given equation is finally reduced to

$$aa'z^2 - \beta = w^2, \tag{2.2}$$

where
$$w = aa'y + \tfrac{1}{2}ab', \tag{2.3}$$

and
$$\beta = a^2a'c' + (\tfrac{1}{4}b^2 - ac)\,aa' - (\tfrac{1}{2}ab')^2.$$

Now, if $z = l$, $w = m$ be a solution of the equation (2.2), another solution will be

$$z = lq \pm mp,$$
$$w = mq \pm aa'lp;$$

where $a'ap^2 + 1 = q^2$. Substituting in (2.1) and (2.3), we get

$$\left.\begin{aligned} x &= -\frac{b}{2a} + \frac{1}{a}(lq \pm mp), \\ y &= -\frac{b'}{2a'} + \frac{1}{aa'}(mq \pm aa'lp). \end{aligned}\right\} \tag{2.4}$$

Now, let $l = ar + \tfrac{1}{2}b$ and $m = aa's + \tfrac{1}{2}ab'$. Substituting in the above expressions, we get the required solution of (2) in the form:

$$\left.\begin{aligned} x &= \frac{1}{2a}(qb \pm pab' - b) + qr \pm pa's, \\ y &= \frac{1}{2a'}(qb' \pm pa'b - b') + qs \pm par; \end{aligned}\right\} \qquad (2.5)$$

where $$aa'p^2 + 1 = q^2,$$

and $$ar^2 + br + c = a's^2 + b's + c'.$$

The form (2.5) shows that having found empirically *one* solution of $ax^2 + bx + c = a'y^2 + b'y + c'$ Bhâskara could find an infinite number of other solutions of it.

Jñânarâja (1503) says:

"If on the other side be present the square as well as the linear power of the unknown together with an absolute term, put it equal to the square of another unknown and then determine the lesser and greater roots. The lesser root will be equal to the first square-root and the greater to the second square-root."

He gives with solution the following illustrative example:

$$3(x^2 + 4x) = y^2 + 4y,$$

or $$(3x + 6)^2 = 3y^2 + 12y + 36.$$

Putting $3x + 6 = z$, where z is the "first square-root" of Jñânarâja, we get

$$z^2 = 3y^2 + 12y^2 + 36,$$

or $$3z^2 = (3y + 6)^2 + 72.$$

Now put $3y + 6 = w$, where w is the "second square-root." Then

$$3z^2 - 72 = w^2.$$

Therefore, by the method of the Square-nature, $z = 18$, $w = 30$. Whence $x = 4$, $y = 8$, is a solution.

(*iii*) *Solution of* $ax^2 + by^2 + c = z^2$

Bhâskara II followed several devices for the solution of the equation

$$ax^2 + by^2 + c = z^2. \qquad (3)$$

In every case his object was to transform the equation into the form of the Square-nature. He says:

"In such cases, where squares of two unknowns with (or without) an absolute number are present, supposing either of them optionally as the *prakṛti*, the rest (of the terms) should be considered as the interpolator. Then the roots should be investigated in the way described before. If there be more equations than one (the process will be especially helpful)."[1]

He then explains further:

"Where on finding the square root of the first side, there remain on the other side squares of two unknowns with or without an absolute number, there consider the square of one of the unknowns as the *prakṛti*; the remainder will then be the interpolator. Then by the rule: 'An optionally chosen number is taken as the lesser root, etc.,'[2] the unknown in the interpolator multiplied by one, etc., and added with one, etc., or not, according to one's own sagacity, should be assumed for the lesser root; then determine the greater root."[3]

There are thus indicated two artifices for solving the equation (3). They are:

(*i*) Set $x = my$; so that equation (3) transforms into

[1] *BBi*, pp. 105f.

[2] The reference is to the rule for solving the Square-nature (*vide supra* p. 144) (*BBi*, p. 33).

[3] *BBi*, p. 106.

$$z^2 = (am^2 + b)y^2 + c$$
$$= \alpha y^2 + c,$$

where $\alpha = am^2 + b$. Hence the required solution of $ax^2 + by^2 + c = z^2$ is

$$x = my = m(rq \pm ps),$$
$$y = rq \pm ps,$$
$$z = sq \pm \alpha pr;$$

where $s^2 = \alpha r^2 + c$ and $q^2 = \alpha p^2 + 1$.

(*ii*) Set $x = my \pm n$; then the equation reduces to

$$z^2 = \alpha y^2 \pm 2amny + \gamma$$

where $\alpha = am^2 + b$ and $\gamma = an^2 + c$.

Completing the square on the right-hand side of this, we get

$$\alpha z^2 - \beta = w^2,$$

where $w = \alpha y \pm amn$ and $\beta = \gamma\alpha - a^2m^2n^2 = a(bn^2 + cm^2) + bc$.

If $z = s$, $w = r$ be a solution of this equation, another solution will be

$$z = sq \pm rp,$$
$$w = rq \pm \alpha sp;$$

where $q^2 = \alpha p^2 + 1$. Hence the solution of $ax^2 + by^2 + c = z^2$ is

$$x = \frac{m}{\alpha}(rq \pm \alpha sp \mp amn) \pm n,$$

$$y = \frac{1}{\alpha}(rq \pm \alpha sp \mp amn),$$

$$z = sq \pm rp;$$

where $q^2 = \alpha p^2 + 1$, $r^2 = \alpha s^2 - \beta$, $\alpha = am^2 + b$ and $\beta = a(bn^2 + cm^2) + bc$.

In working certain problems, Bhâskara II is found to have occasionally followed other artifices also for the solution of the equation (3). For instance:

(*iii*)[1] Set $w^2 = by^2 + c$. Then equation (3) becomes

$$z^2 - w^2 = ax^2.$$

Whence
$$z = \tfrac{1}{2}\left(\frac{a}{m} + m\right)x,$$

and
$$w = \tfrac{1}{2}\left(\frac{a}{m} - m\right)x;$$

where m is an arbitrary number. Therefore

$$x = \frac{2mw}{a - m^2},$$

$$z = \left(\frac{a + m^2}{a - m^2}\right)w.$$

Now, if $y = l$, $w = r$ be a solution of

$$w^2 = by^2 + c,$$

another solution of it will be

$$y = lq \pm pr,$$
$$w = rq \pm blp;$$

where $ap^2 + 1 = q^2$. Therefore, the solution of (3) will be

$$x = \frac{2m}{a - m^2}(rq \pm blp),$$

$$y = lq \pm pr,$$

$$z = \frac{a + m^2}{a - m^2}(rq \pm blp);$$

where $ap^2 + 1 = q^2$ and $bl^2 + c = r^2$.

(*iv*)[2] Suppose $c = 0$; then the equation to be solved will be

[1] See *BBi*, p. 108.

[2] *BBi*, p. 106.

$$ax^2 + by^2 = z^2.$$

In this case set $x = uy$, $z = vy$; so that u, v will be given by

$$au^2 + b = v^2,$$

which can be solved by the method of the Square-nature. Some of these devices were followed also by later Hindu algebraists, for instance, Jñânarâja (1503) and Kamalâkara (1658).[1]

Example from Kamalâkara :[2]

$$7x^2 + 8y^2 = z^2.$$

This is one of a double equation by Bhâskara II.[3]

To solve $ax^2 + by^2 + c = z^2$, Kamalâkara observes:

"In this case, suppose the coefficient of the square of the first unknown as the *prakṛti* and the coefficient of the square of the other unknown together with the absolute number as the interpolator to that. The two roots can thus be determined in several ways."[4]

And again:

"(Suppose) the coefficient of the square of one of the unknowns as the *prakṛti* and the rest comprising two terms, the square of an unknown and an absolute number, as the interpolator. Then assume the value of the lesser root to be equal to the other unknown together with an absolute term."[5]

He seems to have indicated also a slightly different method:

"Or assume the value of the lesser root to be equal to another unknown plus or minus an absolute number

[1] *SiTVi*, xiii. 260-1.
[2] *SiTVi*, xiii. 258.
[3] *BBi*, p. 106.
[4] *SiTVi*, xiii. 264.
[5] *SiTVi*, xiii. 267 f.

and similarly also the value of the greater root. The remaining operations should be performed by the intelligent in the way described by Bhâskara in his algebra."[1]

That is to say, assume

$$x = mw \pm \alpha, \quad z = nw \pm \beta.$$

Substituting in the equation $ax^2 + by^2 + c = z^2$, we get

$$(am^2 - n^2)w^2 \pm 2w\,(am\alpha \mp n\beta) + by^2 + (c + a\alpha^2 - \beta^2) = 0.$$

Putting $\lambda = am^2 - n^2$, $\mu = am\alpha \mp n\beta$, $\nu = c + a\alpha^2 - \beta^2$, this equation can be reduced to

$$-\lambda by^2 + (\mu^2 - \nu\lambda) = u^2,$$

where $u = \lambda w \pm \mu$.

Kamalâkara gives also some other methods which are applicable only in particular cases.

Case i. Suppose that b and c are of different signs.[2] Two sub-cases arise:

(1) *Form* $ax^2 + by^2 - c = z^2$.

First find u, v, says Kamalâkara, such that

$$au^2 - c = v^2.$$

Assuming

$$x = \sqrt{\frac{b}{ac}}\,vy + u,$$

we have

$$ax^2 + by^2 - c = \frac{b}{c}v^2y^2 + 2\sqrt{\frac{ab}{c}}\,uvy + by^2 + (au^2 - c)$$

$$= \frac{b}{c}(au^2 - c)y^2 + 2\sqrt{\frac{ab}{c}}\,uvy + by^2 + v^2$$

[1] *SiTVi*, xiii. 265.

[2] *SiTVi*, xiii. 285-7.

$$= \frac{ab}{c} u^2y^2 + 2\sqrt{\frac{ab}{c}} uvy + v^2$$

$$= \left(\sqrt{\frac{ab}{c}} uy + v\right)^2.$$

Hence $$z = \sqrt{\frac{ab}{c}} uy + v.$$

The following illustrative example and its solution are given :[1]

$$5x^2 + 16y^2 - 20 = z^2.$$

Its solution is

$$x = \tfrac{2}{5}vy + u,$$

$$z = 2uy + v;$$

where $5u^2 - 20 = v^2$. An obvious solution of this equation is given by $u = 3$, $v = 5$. Hence, we get a solution of the given equation as

$$x = 2y + 3,$$

$$z = 6y + 5.$$

Therefore $(x, y) = (5, 1), (7, 2), (9, 3), \ldots\ldots$

(2) *Form* $ax^2 - by^2 + c = z^2$.

In this case first solve

$$au'^2 + c = v'^2.$$

Then the required solution is

$$x = \sqrt{\frac{b}{ac}} v'y + u',$$

$$z = \sqrt{\frac{ab}{c}} u'y + v'.$$

Example from Kamalâkara :[2]

$$5x^2 - 20y^2 + 16 = z^2.$$

[1] *SiTVi*, xiii. 279. [2] *SiTVi*, xiii. 279.

Then

$$x = \tfrac{1}{2}v'y + u',$$
$$z = \tfrac{5}{2}u'y + v';$$

where $5u'^2 + 16 = v'^2$. One solution of this equation is $u' = 2$, $v' = 6$. The corresponding solution of the given equation is

$$x = 3y + 2,$$
$$z = 5y + 6.$$

Therefore $(x, y) = (5, 1), (8, 2), (11, 3)$, etc.

Case ii. Let the two terms of the interpolator be of the same sign and positive.

Example from Kamalâkara:[1]

$$5x^2 + 8y^2 + 23 = z^2.$$

Assume arbitrarily a value of x or y and then find the other by the method of the Square-nature.[2]

(*iv*) *Solution of* $a^2x^2 + by^2 + c = z^2$

Let the coefficient of x^2 (or y^2) be a square number. The equation is of the form

$$a^2x^2 + by^2 + c = z^2.$$

For this case Bhâskara II observes :

"If the *prakṛti* is a square, then obtain the roots by the rule: 'The interpolator divided by an optional number is set down at two places, etc.' "[3]

Thus, according to Bhâskara II, the solution of the above equation is

$$x = \frac{1}{2a}\left(\frac{by^2 + c}{m}\right) - m,$$

[1] *SiTVi*, xiii. 296. [2] *SiTVi*, xiii. 298.
[3] *BBi*, p. 106.

$$z = \tfrac{1}{2}\left(\frac{by^2 + c}{m} + m\right);$$

where m is an arbitrary number.

Kamalâkara divides equations of this form into two classes according as c is or is not a square.[1]

(1) Let c be a square ($= d^2$, say). That is to say, we have to solve

$$a^2x^2 + by^2 + d^2 = z^2.$$

The solution of this particular case, says Kamalâkara, is given by

$$x = \frac{b}{2ad}y^2.$$

For, with this value, we have

$$z^2 = \frac{b^2y^4}{4d^2} + by^2 + d^2$$

$$= \frac{1}{4d^2}(by^2 + 2d^2)^2.$$

Hence $$z = \frac{by^2}{2d} + d.$$

(2) When c is not a square, Kamalâkara first finds α, β such that

$$\alpha^2 + c = \beta^2.$$

He next obtains n such that the value of

$$\frac{a\beta n - b/2}{a^2 . \alpha/a}$$

is also n; and then says that

$$x = ny^2 + \frac{\alpha}{a};$$

[1] *Vide* his gloss on *SiṪVi*, xiii. 275.

whence will follow the value of z.

Since $$\frac{a\beta n - b/2}{a\alpha} = n,$$

we get $$n = \frac{b}{2a(\beta - \alpha)};$$

so that $$x = \frac{by^2}{2a(\beta - \alpha)} + \frac{\alpha}{a}.$$

Therefore $$z^2 = a^2x^2 + by^2 + c$$

$$= \frac{b^2y^4}{4(\beta - \alpha)^2} + \alpha^2 + \frac{\alpha by^2}{(\beta - \alpha)} + by^2 + c$$

$$= \frac{b^2y^4}{4(\beta - \alpha)^2} + \frac{\beta by^2}{(\beta - \alpha)} + \beta^2$$

$$= \left\{\frac{by^2}{2(\beta - \alpha)} + \beta\right\}^2.$$

Hence $$z = \frac{by^2}{2(\beta - \alpha)} + \beta.$$

Example from Kamalâkara :

$$4x^2 + 48y^2 + 20 = z^2.$$

Since $4^2 + 20 = 6^2$, and the solution of

$$\frac{12n - 24}{8} = n$$

is $n = 6$, we get the required solution of the given equation as

$$x = 6y^2 + 2,$$
$$z = 12y^2 + 6.$$

It may be noted that the solution stated by Kamalâkara follows easily from that of Bhâskara II, on putting therein $m = \beta - \alpha$, where $\alpha^2 + c = \beta^2$.

In particular, if we put $c = 0$ and $a = 1$, Bhâskara's solution reduces to

$$x = \tfrac{1}{2}\left(\frac{b}{m}y^2 - m\right),$$

$$z = \tfrac{1}{2}\left(\frac{b}{m}y^2 + m\right);$$

where m is arbitrary.

If $b = \alpha\beta$, taking $m = 2\beta p^2$, we easily arrive at

$$x = \alpha q^2 - \beta p^2,$$
$$y = 2pq,$$
$$z = \alpha q^2 + \beta p^2,$$

where p, q are arbitrary integers, as the solution in positive integers of $z^2 = x^2 + by^2$. This solution was given by A. Desboves (1879).[1]

Taking $m = 2\nu^2$, we can derive Matsunago's (*c.* 1735) solution of $z^2 = x^2 + by^2$, *viz.*,

$$x = b\mu^2 - \nu^2,$$
$$y = 2\mu\nu,$$
$$z = b\mu^2 + \nu^2;$$

where μ, ν are arbitrary integers.

(*v*) *Solution of* $ax^2 + bxy + cy^2 = z^2$

For the solution of the equation

$$ax^2 + bxy + cy^2 = z^2 \qquad (5)$$

Bhâskara II lays down the following rule:

"When there are squares of two unknowns together with their product, having extracted the square-root of one part, it should be put equal to half the difference of the remaining part divided by an optional number

[1] Dickson, *Numbers*, II, p. 405.

and the optional number."[1]

It has been again elucidated thus:

"Where there exists also the product of the unknowns (in addition to their squares), by the rule, 'when there are squares etc.,' the square-root of as much portion of it as affords a root, should be extracted. The remaining portion, divided by an optional number and then diminished by that optional number and halved, should be put equal to that square-root."[2]

The above rule, in fact, contemplates a particular case of equation (5) in which a or c is a square number.

(*i*) Suppose $a = p^2$. The equation to be solved is then

$$p^2x^2 + bxy + cy^2 = z^2. \qquad (\ldots)$$

Therefore $$\left(px + \frac{by}{2p}\right)^2 + y^2\left(c - \frac{b^2}{4p^2}\right) = z^2.$$

Putting $px + \dfrac{by}{2p} = w$, we get

$$z^2 - w^2 = y^2\left(c - \frac{b^2}{4p^2}\right).$$

Whence $$z - w = \lambda,$$

$$z + w = \frac{y^2}{\lambda}\left(c - \frac{b^2}{4p^2}\right),$$

where λ is an arbitrary rational number. So

$$w = \tfrac{1}{2}\left\{\frac{y^2}{\lambda}\left(c - \frac{b^2}{4p^2}\right) - \lambda\right\},$$

as stated in the rule. Therefore,

$$x = \frac{1}{2p}\left\{\frac{y^2}{\lambda}\left(c - \frac{b^2}{4p^2}\right) - \lambda\right\} - \frac{by}{2p^2},$$

[1] Y. Mikami, *The Development of Mathematics in China and Japan*, Leipzig, 1913, p. 231.

[2] *BBi*, p. 106.

and $$z = \tfrac{1}{2}\left\{\frac{y^2}{\lambda}\left(c - \frac{b^2}{4p^2}\right) + \lambda\right\}.$$

Now, if we suppose $y = m/n$, where m, n are arbitrary integers, we get the solution of (5.1) as

$$x = \frac{1}{8\lambda p^2 n^2}\{m^2(4cp^2 - b^2) - 4\lambda^2 p^2 n^2 - 4\lambda bmn\},$$

$$y = \frac{m}{n},$$

$$z = \frac{1}{8\lambda p^2 n^2}\{m^2(4cp^2 - b^2) + 4\lambda^2 p^2 n^2\}.$$

Since the given equation is homogeneous, any multiple of these values of x, y, z will also be its solution. Therefore, multiplying by $8\lambda p^2 n^2$, we get the following solution of the equation $p^2x^2 + bxy + cy^2 = z^2$ in integers :

$$\left.\begin{aligned} x &= m^2(4cp^2 - b^2) - 4\lambda^2 p^2 n^2 - 4\lambda bmn,\\ y &= 8\lambda mnp^2,\\ z &= m^2(4cp^2 - b^2) + 4\lambda^2 p^2 n^2, \end{aligned}\right\} \quad (5.2)$$

where m, n, are arbitrary integers.

In particular, putting $a = b = c = 1$, and $\lambda = p = 1$ in (5.2), we get

$$x = 3m^2 - 4n(n + m),$$
$$y = 8mn,$$
$$z = 3m^2 + 4n^2,$$

as the solution of the equation

$$x^2 + xy + y^2 = z^2.$$

Dividing out by $8n$, the above solution can be put into the form

$$x = \tfrac{1}{2}\left(\frac{3m^2}{4n} - n - m\right),$$

$$y = m,$$

$$z = \frac{1}{8n}(3m^2 + 4n^2);$$

as has been stated by Nârâyaṇa:

"An arbitrary number is the first. Its square less by its (square's) one-fourth, is divided by an optional number and then diminished by the latter and also by the first. Half the remainder is the second number. The sum of their squares together with their product is a square."[1]

It is noteworthy that in practice Nârâyaṇa approves of only integral solutions of his equation. For instance, he says:

"'Any arbitrary number is the first.' Suppose it to be 12. Then with the optional number unity, are obtained the numbers (12, 95/2). For *integral values*, they are doubled (24, 95). With the optional number 2, are obtained (12, 20). It being possible, these are reduced by the common factor 4 to (3, 5). In this way, owing to the varieties of the optional number, an infinite number of solutions can be obtained."[2]

(*ii*) If neither a nor c be a square, the solution can be obtained thus:

Multiplying both sides of the equation (5) by a and then completing a square on the left-hand side, the equation transforms into

$$(ax + \tfrac{1}{2}by)^2 + (ac - \tfrac{1}{4}b^2)y^2 = az^2.$$

Putting $ax + \frac{1}{2}by = w$ and $\beta = \frac{1}{4}(b^2 - 4ac)$,

we get $w^2 = az^2 + \beta y^2.$ (5·3)

[1] *GK*, i. 55.

[2] See the example in illustration of the same.

The method of the solution of an equation of this form, according to Bhâskara II, has been described before.

Assume $w = vy$, $z = uy$; so that the values of u, v will be given by

$$v^2 = au^2 + \beta. \tag{5.4}$$

If $u = m$, $v = n$ be a solution of (5.4), another solution will be

$$u = mq \pm pn,$$

$$v = nq \pm amp;$$

where $ap^2 + 1 = q^2$. Therefore, a solution of (5) is

$$x = \frac{y}{2a}\{2(nq \pm amp) - b\},$$

$$z = y(mq \pm pn);$$

where $ap^2 + 1 = q^2$ and $am^2 + \beta = n^2$.

Put $n = ar + \frac{1}{2}b$ and $y = \frac{s}{t}$; then we have

$$x = \frac{s}{2at}\{q(2ar + b) \pm 2amp - b\},$$

$$y = \frac{s}{t},$$

$$z = \frac{s}{2t}\{2mq \pm p(2ar + b)\}.$$

Multiplying by $2at$, we get the following solution of $ax^2 + bxy + cy^2 = z^2$ in integers :

$$x = s\{q(2ar + b) \pm 2amp - b\},$$

$$y = 2as,$$

$$z = as\{2mq \pm p(2ar + b)\};$$

where $ap^2 + 1 = q^2$ and $m^2 = ar^2 + br + c$.

20. RATIONAL TRIANGLES

Rational Right Triangles: Early Solutions. The earliest Hindu solutions of the equation

$$x^2 + y^2 = z^2 \tag{1}$$

are found in the *Śulba*. Baudhâyana (*c.* 800 B.C.), Âpastamba and Kâtyâyana (*c.* 500 B.C.)[1] give a method for the transformation of a rectangle into a square, which is the equivalent of the algebraical identity

$$mn = \left(m - \frac{m-n}{2}\right)^2 - \left(\frac{m-n}{2}\right)^2,$$

where m, n are any two arbitrary numbers. Thus we get

$$(\sqrt{mn})^2 + \left(\frac{m-n}{2}\right)^2 = \left(\frac{m+n}{2}\right)^2.$$

Substituting p^2, q^2 for m, n respectively, in order to eliminate the irrational quantities, we get

$$p^2q^2 + \left(\frac{p^2-q^2}{2}\right)^2 = \left(\frac{p^2+q^2}{2}\right)^2,$$

which gives a rational solution of (1).

For finding a square equal to the sum of a number of other squares of the same size, Kâtyâyana gives a very elegant and simple method which furnishes us with another solution of the rational right triangle. Kâtyâyana says:

"As many squares (of equal size) as you wish to combine into one, the transverse line will be (equal to) one less than that; twice a side will be (equal to) one more than that; (thus) form (an isosceles) triangle. Its arrow (*i.e.*, altitude) will do that."[2]

[1] *BŚl*, i. 58; *ÂpŚl*, ii. 7; *KŚl*, iii. 2. For details of the construction see Datta, *Śulba*, pp. 83f, 178f.

[2] *KŚl*, vi. 5; Compare also its *Pariśiṣṭa*, verses 40-1.

Thus for combining n squares of sides a each, we form the isosceles triangle ABC, such that $AB=AC=(n+1)a/2$,

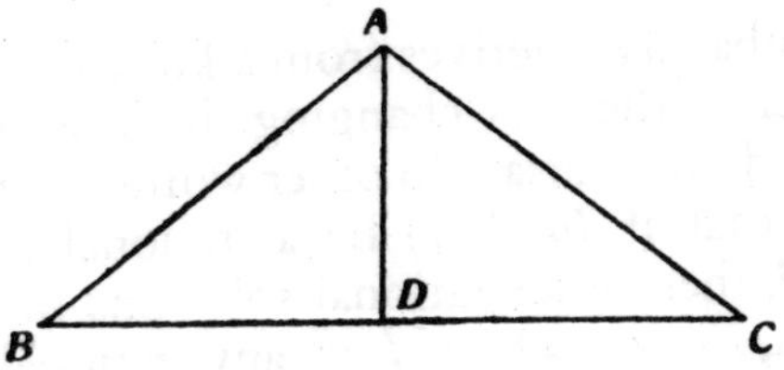

Fig. 2

and $BC=(n-1)a$. Then $AD^2=na^2$. This gives the formula

$$a^2(\sqrt{n})^2+a^2\left(\frac{n-1}{2}\right)^2=a^2\left(\frac{n+1}{2}\right)^2.$$

Putting m^2 for n in order to make the sides of the right-angled triangle free from the radical, we have

$$m^2a^2+\left(\frac{m^2-1}{2}\right)^2a^2=\left(\frac{m^2+1}{2}\right)a^2,$$

which gives a rational solution of (1).

Tacit assumption of the following further generalisation is met with in certain constructions described by Āpastamba :[1]

If the sides of a rational right triangle be increased by any rational multiple of them, the resulting figure will be a right triangle.

In particular, he notes

$$3^2+4^2=5^2,$$
$$(3+3\cdot3)^2+(4+4\cdot3)^2=(5+5\cdot3)^2,$$
$$(3+3\cdot4)^2+(4+4\cdot4)^2=(5+5\cdot4)^2;$$

[1] *ĀpŚl*, v. 3, 4. Also compare Datta, *Śulba*, pp. 65f

$$5^2 + 12^2 = 13^2,$$

$$(5 + 5.2)^2 + (12 + 12.2)^2 = (13 + 13.2)^2.$$

Âpastamba also derives from a known right-angled triangle several others by changing the unit of measure of its sides and vice versa.[1] In other words, he recognised the principle that if (α, β, γ) be a rational solution of $x^2 + y^2 = z^2$, then other rational solutions of it will be given by $(l\alpha, l\beta, l\gamma)$, where l is any rational number. This is clearly in evidence in the formula of Kâtyâyana in which a is any quantity. It is now known that all rational solutions of $x^2 + y^2 = z^2$ can be obtained without duplication in this way.

Later Rational Solutions. Brahmagupta (628) says:

"The square of the optional (*iṣṭa*) side is divided and then diminished by an optional number; half the result is the upright, and that increased by the optional number gives the hypotenuse of a rectangle."[2]

In other words, if m, n be any two rational numbers, then the sides of a right triangle will be

$$m, \tfrac{1}{2}\left(\frac{m^2}{n} - n\right), \tfrac{1}{2}\left(\frac{m^2}{n} + n\right).$$

The Sanskrit word *iṣṭa* can be interpreted as implying "given" as well as "optional". With the former meaning the rule will state how to find rational right triangles having a given leg. Such is, in fact, the interpretation which has been given to a similar rule of Bhâskara II.[3]

[1] Datta, *Sulba*, p. 179. [2] *BrSpSi*, xii. 35.

[3] *Vide infra* p. 211; H. T. Colebrooke, *Algebra with Arithmetic and Mensuration from the Sanscrit of Brahmegupta and Bhascara*, London, 1817, (referred to hereafter as, Colebrooke, *Hindu Algebra*), p. 61 footnote.

A similar rule is given by Śrîpati (1039):

"Any optional number is the side; the square of that divided and then diminished by an optional number and halved is the upright; that added with the previous divisor is the hypotenuse of a right-angled triangle. For, so it has been explained by the learned in the matter of the rules of geometry."[1]

Karavindasvamî a commentator of the *Âpastamba Sulba*, finds the solution

$$m, \left(\frac{n^2 + 2n}{2n + 2}\right) m, \left(\frac{n^2 + 2n + 2}{2n + 2}\right) m,$$

by generalising a rule of the *Śulba*.[2]

Integral Solutions. Brahmagupta was the first to give a solution of the equation $x^2 + y^2 = z^2$ in integers. It is

$$m^2 - n^2, \; 2mn, \; m^2 + n^2,$$

m, n being any two unequal integers.[3]

Mahâvîra (850) says:

"The difference of the squares (of two elements) is the upright, twice their product is the base and the sum of their squares is the diagonal of a *generated* rectangle."[4]

He has re-stated it thus:

"The product of the sum and difference of the elements is the upright. The *saṅkramaṇa*[5] of their squares gives the base and the diagonal. In the operation of generating (geometrical figures), this is the process."[6]

[1] *SiŚe*, xiii. 41.
[2] *ÂpŚl*, 1.2 (*Com.*); also see Datta, *Śulba*, pp. 14-16.
[3] *BrSpSi*, xii. 33; *vide infra*, p. 222. [4] *GSS*, vii. 90½.
[5] For the definition of this term see pp. 43f.
[6] *GSS*, vii. 93½.

Bhâskara II (1150) writes:

"Twice the product of two optional numbers is the upright; the difference of their squares is the side; and the sum of their squares is the hypotenuse. (Each of these quantities is) rational (and integral)."[1]

It has been stated before that the early Hindus recognised that fresh rational right triangles can be derived from a known one by multiplying or dividing its sides by any rational number. The same principle has been used by Mahâvîra and Bhâskara II in their treatment of the solution of rational triangles and quadrilaterals. Gaṇeśa (1545) expressly states:

"If the upright, base and hypotenuse of a rational right-angled triangle be multiplied by any arbitrary rational number, there will be produced another right-angled triangle with rational sides."

Hence the most general solution of $x^2 + y^2 = z^2$ in integers is

$$(m^2 - n^2)l, \; 2mnl, \; (m^2 + n^2)l$$

where m, n, l are integral numbers.

Mahâvîra's Definitions. A triangle or a quadrilateral whose sides, altitudes and other dimensions can be expressed in terms of *rational* numbers is called *janya* (meaning generated, formed or that which is generated or formed) by Mahâvîra.[2] Numbers which

[1] *L*, p. 36.

[2] *GSS*, introductory line to vii. 90½. The section of Mahâvîra's work devoted to the treatment of rational triangles and quadrilaterals bears the sub-title *janya-vyavahâra* (*janya* operation) and it begins as "Hereafter we shall give out the *janya* operations in calculations relating to measurement of areas." Mahâvîra's treatment of the subject has been explained fully by Bibhutibhusan Datta in a paper entitled: "On Mahâvîra's solution of rational triangles and quadrilaterals." *BCMS*, XX, 1928-9, pp. 267-294.

are employed in forming a particular figure are called its *bîja-saṁkhyâ* (element-numbers) or simply *bîja* (element or seed). For instance, Mahâvîra has said: "Forming O friend! the generated figure from the *bîja* 2, 3,"[1] "forming another from half the base of the figure (rectangle) from the *bîja* 2, 3,"[2] etc. Thus, according to Mahâvîra, "forming a rectangle from the *bîja* m, n" means taking a rectangle with the upright, base and diagonal as $m^2 - n^2$, $2mn$, $m^2 + n^2$ respectively. It is noteworthy that Mahâvîra's mode of expression in this respect very closely resembles that of Diophantus who also says, "Forming now a right-angled triangle from 7, 4," meaning "taking a right-angled triangle with sides $7^2 - 4^2$, $2.7.4$, $7^2 + 4^2$ or 33, 56, 65."[3] It should also be noted that Mahâvîra never speaks of "right-angled triangle." What Diophantus called "forming a right-angled triangle from m, n," Mahâvîra calls "forming a longish quadrilateral or rectangle from m, n."

Right Triangles having a Given Side. In the *Śulba* we find an attempt to find rational right triangles having a given side, that is, rational solutions of

$$x^2 + a^2 = z^2.$$

In particular, we find mention of two such right triangles having a common side a, *viz.*, $(a, 3a/4, 5a/4)$ and $(a, 5a/12, 13a/12)$.[4] The principle underlying these solutions will be easily detected to be that of the reduction of the sides of any rational right triangle in the ratio of the given side to its corresponding

[1] "Bîje dve trîṇi sakhe kṣetre janye tu saṁsthâpya"—*GSS*, vii. 92½.

[2] "He dvitribîjakasya kṣetrabhujârdhena cânyamutthâpya"—*GSS*, vii. 111½.

[3] *Arithmetica*, Book III, 19; T. L. Heath, *Diophantus of Alexandria*, p. 167.

[4] Datta, *Śulba*, p. 180.

side. This principle of finding rational right triangles having a given side has been followed explicitly by Mahâvîra (850).[1]

It has been stated before that one rule of Brahmagupta[2] can be interpreted as giving rational solutions of $x^2 + a^2 = z^2$ as

$$a, \ \tfrac{1}{2}\left(\frac{a^2}{n} - n\right), \ \tfrac{1}{2}\left(\frac{a^2}{n} + n\right),$$

where n is any rational number. In fact, he has used this solution in finding rational isosceles triangles having a given altitude.[3] This solution has been expressly stated by Mahâvîra (850). He says:

"The *saṅkramaṇa* between any optional divisor of the square of the given upright or the base and the (respective) quotient gives the diagonal and the base (or upright)."[4]

He has restated the solution thus:

"The *saṅkramaṇa* between any (rational) divisor of the upright and the quotient gives the elements; or any (rational) divisor of half the side and the quotient are the elements."[5]

The right triangles formed according to the first half of this rule are:[6]

$$a, \ \tfrac{1}{2}\left(\frac{a^2}{p^2} - p^2\right), \ \tfrac{1}{2}\left(\frac{a^2}{p^2} + p^2\right),$$

[1] *Vide infra*, p. 213

[2] *Vide supra*, p. 206.

[3] *Vide infra*, p. 223

[4] *GSS*, vii. 97½.

[5] *GSS*, vii. 95½.

[6] The "elements" here are $\frac{1}{2}(a/p + p)$, $\frac{1}{2}(a/p - p)$, where p is an optional number.

and those according to the second half are :[1]

$$\frac{a^2}{4q^2} - q^2,\ a,\ \frac{a^2}{4q^2} + q^2.$$

Bhâskara II gives two solutions one of which is the same as that of Brahmagupta. He says :

"The side is given : from that multiplied by twice an optional number and divided by the square of that optional number minus unity, is obtained the upright ; this again multiplied by the optional number and diminished by the given side becomes the hypotenuse. This triangle is a right-angled triangle.

"Or the side is given : its square divided by an optional number is put down at two places ; the optional number is subtracted (at one place) and added (at another) and then halved ; these results are the upright and the hypotenuse. Similarly from the given upright can be obtained the side and the hypotenuse."[2]

That is to say, the two solutions are

$$a,\ \frac{2na}{n^2 - 1},\ n\left(\frac{2na}{n^2 - 1}\right) - a,$$

and $$a,\ \tfrac{1}{2}\left(\frac{a^2}{n} - n\right),\ \tfrac{1}{2}\left(\frac{a^2}{n} + n\right).$$

Bhâskara II illustrates this by finding four right triangles having a side equal to 12, *viz*., (12, 35, 37), (12, 16, 20), (12, 9, 15) and (12, 5, 13).[3]

The *rationale* of the first solution has been given by Sûryadâsa (1538) thus : Starting with the rational right triangle $n^2 - 1$, $2n$, $n^2 + 1$, he observes that if x, y, z

[1] The "elements" here are q, $a/2q$, where q is an optional number.

[2] *L*, p. 34.

[3] *L*, pp. 34f.

be the corresponding sides of another right triangle, then

$$\frac{x}{n^2-1} = \frac{y}{2n} = \frac{z}{n^2+1} = k \text{ (say)}.$$

Hence $\quad x = k(n^2-1), \; y = 2nk, \; z = k(n^2+1).$

Therefore $\quad x + z = 2kn^2 = ny.$

If now we have $x = a$, then

$$k = \frac{a}{n^2-1}.$$

Hence $$y = \frac{2na}{n^2-1},$$

and $$z = ny - a = n\left(\frac{2na}{n^2-1}\right) - a.$$

The second rule has been demonstrated by Sûryadâsa, Gaṇeśa and Raṅganâtha thus :

Since $\quad x^2 + a^2 = z^2,$

we have $\quad a^2 = z^2 - x^2 = (z - x)(z + x).$

Assume $z - x = n$, where n is any rational number ; then

$$z + x = \frac{a^2}{n}.$$

$$\therefore \quad z = \tfrac{1}{2}\left(\frac{a^2}{n} + n\right), \; x = \tfrac{1}{2}\left(\frac{a^2}{n} - n\right).$$

Generalising the method of the *Âpastamba Sulba* the commentators obtained the solution[1]

$$a, \left(\frac{m^2+2m}{2m+2}\right)a, \left(\frac{m^2+2m+2}{2m+2}\right)a.$$

[1] Datta, *Śulba*, p. 16.

Right Triangles having a Given Hypotenuse. For finding all rational right triangles having a given hypotenuse (c), that is, for rational solutions of

$$x^2 + y^2 = c^2,$$

Mahâvîra gives three rules. The first rule is:

"The square-root of half the sum and difference of the diagonal and the square of an optional number are they (the elements)."[1]

In other words, the required solution will be obtained from the "elements" $\sqrt{(c+p^2)/2}$ and $\sqrt{(c-p^2)/2}$, where p is any rational number. Hence the solution is

$$p^2, \sqrt{c^2 - p^4}, \quad c.$$

The second rule is:

"Or the square-root of the difference of the squares of the diagonal and of an optional number, and that optional number are the upright and the base."[2]

That is, the solution is

$$p, \sqrt{c^2 - p^2}, \quad c.$$

These solutions are defective in the sense that $\sqrt{c^2 - p^4}$ or $\sqrt{c^2 - p^2}$ might not be rational unless p is suitably chosen. Mahâvîra's third rule is of greater importance. He says:

"Each of the various figures (rectangles) that can be formed from the elements are put down; by its diagonal is divided the given diagonal. The perpendicular, base and the diagonal (of this figure) multiplied by this quotient (give rise to the corresponding sides of the figure having the given hypotenuse)."[3]

[1] *GSS*, vii. 95½.
[2] *GSS*, vii. 97½.
[3] *GSS*, vii. 122½.

Thus having obtained the general solution of the rational right triangle, *viz.*, $m^2 - n^2$, $2mn$, $m^2 + n^2$, Mahâvîra reduces it in the ratio $c/(m^2 + n^2)$, so that all rational right triangles having a given hypotenuse c will be given by

$$\left(\frac{m^2 - n^2}{m^2 + n^2}\right)c, \ \left(\frac{2mn}{m^2 + n^2}\right)c, \ c.$$

By way of illustration Mahâvîra finds four rectangles (39, 52), (25, 60), (33, 56) and (16, 63) having the same diagonal 65.[1]

This method was later on rediscovered in Europe by Leonardo Fibonacci of Pisa (1202) and Vieta. It has been pointed out before that the origin of the method can be traced to the *Sulba*.

Bhâskara II (1150) says:

"From the given hypotenuse multiplied by an optional number and doubled and then divided by the square of the optional number added to unity, is obtained the upright; this is again multiplied by the optional number; the difference between that (product) and the given hypotenuse is the side.

"Or divide twice the hypotenuse by the square of an optional number added to unity. The hypotenuse minus the quotient is the upright and the quotient multiplied by the optional number is the side."[2]

Thus, according to the above, the sides of a right-angled triangle whose hypotenuse is c are:

$$\frac{2mc}{m^2 + 1}, \ m\left(\frac{2mc}{m^2 + 1}\right) - c, \ c;$$

or

$$\frac{2mc}{m^2 + 1}, \ c - \frac{2c}{m^2 + 1}, \ c.$$

[1] *GSS*, vii. 123-124½.

[2] *L*, pp. 35, 36.

By way of illustration Bhâskara II finds two right triangles (51, 68) and (40, 75) having the same hypotenuse 85.[1]

Sûryadâsa demonstrates the above substantially thus:

If (x, y, z) be the sides of the right triangle, we have

$$\frac{x}{m^2-1} = \frac{y}{2m} = \frac{z}{m^2+1} = k \text{ (say)},$$

where m is any rational integer. Then

$$x = k(m^2-1),\ y = 2mk,\ z = k(m^2+1).$$

Therefore $$x + z = 2km^2 = my.$$

Since z is given to be equal to c, we have

$$k = \frac{c}{m^2+1}.$$

Hence $$y = \frac{2mc}{m^2+1},$$

and $$x = my - z = m\left(\frac{2mc}{m^2+1}\right) - c.$$

Problems Involving Areas and Sides. Mahâvîra proposes to find rational rectangles (or squares) in which the area will be *numerically* (*saṁkhyayâ*) equal to any multiple or submultiple of a side, diagonal or perimeter, or of any linear combination of two or more of them. Expressed symbolically, the problem is to solve

$$\left.\begin{aligned} x^2 + y^2 &= z^2, \\ mx + ny + pz &= rxy; \end{aligned}\right\} \qquad (1)$$

m, n, p, r being any rational numbers $(r \neq 0)$. For the solution of this problem he gives the following rule:

[1] *L*, pp. 35f.

"Divide the sides (or their sum) of any generated square or other figure as multiplied by their respective given multiples by the area of that figure taken into its given multiple. The sides of that figure multiplied by this quotient will be the sides of the (required) square or other figure."[1]

That is to say, starting with any rational solution of

$$x'^2 + y'^2 = z'^2, \qquad (2)$$

we shall have to calculate the value of

$$mx' + ny' + pz' = Q, \text{ say}. \qquad (3)$$

Then the required solution of (1) will be obtained by reducing the values of x', y', z' in the ratio of $Q/rx'y'$. Thus

$$\left.\begin{aligned} x &= x'Q/rx'y' = Q/ry', \\ y &= y'Q/rx'y' = Q/rx', \\ z &= z'Q/rx'y'. \end{aligned}\right\} \qquad (4)$$

Mahâvîra gives several illustrative examples some of which are very interesting:

"In a rectangle the area is (numerically) equal to the perimeter; in another rectangle the area is (numerically) equal to the diagonal. What are the sides (in each of these cases)?"[2]

Algebraically, we shall have to solve

$$\left.\begin{aligned} x^2 + y^2 &= z^2, \\ 2(x + y) &= xy, \end{aligned}\right\} \qquad (1.1)$$

and

$$\left.\begin{aligned} x^2 + y^2 &= z^2, \\ xy &= z. \end{aligned}\right\} \qquad (1.2)$$

Starting with the solution $s^2 - t^2$, $2st$, $s^2 + t^2$ of (2) and putting $m = n = 2$, $p = 0$, $r = 1$ in (4), we get

[1] *GSS*, vii. 112½. [2] *GSS*, vii. 115½.

the solution of (1.1) as

$$\frac{2(s^2-t^2)+4st}{2st},\ \frac{2(s^2-t^2)+4st}{s^2-t^2},$$
$$\left\{\frac{2(s^2-t^2)+4st}{2st\,(s^2-t^2)}\right\}(s^2+t^2).$$

And putting $m = n = 0$, $p = r = 1$, in (4), we have the solution of (1.2) as

$$\frac{s^2+t^2}{2st},\ \frac{s^2+t^2}{s^2-t^2},\ \frac{(s^2+t^2)^2}{2st(s^2-t^2)}.$$

Bhâskara II solves a problem similar to the second one above:

Find a right triangle whose area equals the hypotenuse.[1]

He starts with the rational right triangle ($3x$, $4x$, $5x$); then by the condition, area = hypotenuse, finds the value $x = 5/6$. So that a right triangle of the required type is ($5/2$, $10/3$, $25/6$). He then observes: "In like manner, by virtue of various assumptions, other right triangles can also be found."[2] The general solution in this case is

$$\frac{s^2+t^2}{st},\ \frac{2(s^2+t^2)}{s^2-t^2},\ \frac{(s^2+t^2)^2}{st(s^2-t^2)}.$$

Another example of Mahâvîra runs as follows:

"(Find) a rectangle of which twice the diagonal, thrice the base, four times the upright and twice the perimeter are together equal to the area (numerically)."[3]

Problems Involving Sides but not Areas. Mahâvîra also obtained right triangles whose sides multiplied

[1] *BBi*, p. 56.
[2] "Evamiṣṭavaśâdanye'pi"—*BBi*, p. 56.
[3] *GSS*, vii. 117½.

by arbitrary rational numbers have a given sum. Algebraically, the problems require the solution of

$$\left.\begin{aligned} x^2+y^2&=z^2,\\ rx+sy+tz&=A\,; \end{aligned}\right\}$$

where r, s, t, A are known rational numbers. His method of solution is the same as that described above. Starting with the general solution of

$$x'^2+y'^2=z'^2$$

we are asked to calculate the value

$$rx'+sy'+tz'=P,\ \text{say}.$$

Then, says Mahâvîra, the required solution is

$$x=x'A/P,\quad y=y'A/P,\quad z=z'A/P.$$

One illustrative problem given by Mahâvîra is :

"The perimeter of a rectangle is unity. Tell me quickly, after calculating, what are its base and upright."[1]

Starting with the rectangle m^2-n^2, $2mn$, m^2+n^2, we have in this case $P=2(m^2-n^2+2mn)$. Hence all rectangles having the same perimeter unity will be given by

$$\frac{m^2-n^2}{2(m^2-n^2+2mn)},\quad \frac{mn}{m^2-n^2+2mn};$$

m, n being any rational numbers.

The isoperimetric right triangles will be given by

$$\left(\frac{m-n}{2m}\right)p,\quad \frac{np}{m+n},\quad \left\{\frac{m^2+n^2}{2m(m+n)}\right\}p\,;$$

where p is the given perimeter.

Another example is :

"(Find) a rectangle in which twice the diagonal, thrice the base, four times the upright and the perimeter together equal unity."[2]

[1] *GSS*, vii. 118½. [2] *GSS*, vii. 119½.

Pairs of Rectangles. Mahâvîra found "pairs of rectangles such that

(*i*) their perimeters are equal but the area of one is double that of the other, or

(*ii*) their areas are equal but the perimeter of one is double that of the other, or

(*iii*) the perimeter of one is double that of the other and the area of the latter is double that of the former."

These are particular cases of the following general problem contemplated in his rule :

To find (x, y) and (u, v) representing the base and upright respectively of two rectangles which are related, such that

$$\left.\begin{aligned} 2m(x+y) &= 2n(u+v), \\ pxy &= quv; \end{aligned}\right\} \qquad (A)$$

where m, n, p, q are known integers.

His rule for the solution of this general problem is :

"Divide the greater multiples of the area and the perimeter by the (respective) smaller ones. The square of the product of these ratios multiplied by an optional number is the upright of one rectangle. That diminished by unity will be its base, when the areas are equal. Otherwise, multiply the bigger ratio of the areas by that optional number and subtract unity ; three times the upright diminished by this (difference) will be the base. The upright and base of the other rectangle should be obtained from its area and perimeter (thus determined) with the help of the rule, 'From the square of half the perimeter, etc.,' described before."[1]

[1] *GSS*, vii. 131½-133. The reference in the concluding line is to rule vii. 129½.

In other words, to solve (A), assume

$$y = s\{(\text{ratio of perimeters})(\text{ratio of areas})\}^2, \qquad (1)$$

and $x = y - 1$, if $p = q$, (2)

or $x = 3[y - \{s\,(\text{ratio of areas}) - 1\}]$, if $p \neq q$, (2′)

where s is an arbitrary number, and the ratios are to be so presented as always to remain greater than or equal to unity.

Let $m \geqslant n$, $q \geqslant p$. Then we shall have to assume

$$\left.\begin{aligned} y &= s\frac{m^2q^2}{n^2p^2}, \\ x &= 3\left(s\frac{m^2q^2}{n^2p^2} - s\frac{q}{p} + 1\right). \end{aligned}\right\} \qquad (3)$$

Substituting these values in (A), we get

$$u + v = \frac{m}{n}\left(4s\frac{m^2q^2}{n^2p^2} - 3s\frac{q}{p} + 3\right), \qquad (4)$$

$$uv = 3s\frac{m^2q}{n^2p}\left(s\frac{m^2q^2}{n^2p^2} - s\frac{q}{p} + 1\right).$$

Then

$$(u - v)^2 = \frac{m^2}{n^2}\left\{\left(4s\frac{m^2q^2}{n^2p^2} - \frac{9sq}{2p} + 3\right)^2 + \frac{3sq}{4p}\left(\frac{sq}{p} - 4\right)\right\}.$$

Now, if the arbitrary multiplier s be chosen such that

$$\frac{sq}{p} = 4, \qquad (5)$$

we have

$$u - v = \frac{m}{n}\left(4s\frac{m^2q^2}{n^2p^2} - \frac{9sq}{2p} + 3\right). \qquad (6)$$

From (4) and (6) we get

$$\left.\begin{aligned} u &= \frac{m}{n}\left(4s\frac{m^2q^2}{n^2p^2} - \frac{15sq}{4p} + 3\right), \\ v &= \frac{3smq}{4np}. \end{aligned}\right\} \qquad (7)$$

Substituting the value of s from (5) in (3) and (7) we have finally the solution of (A), when $m \geqslant n$, $q \geqslant p$, as

$$\left.\begin{aligned} y &= 4\frac{m^2q}{n^2p}, & v &= 3\frac{m}{n}, \\ x &= 3\left(4\frac{m^2q}{n^2p} - 3\right), & u &= 4\frac{m}{n}\left(4\frac{m^2q}{n^2p} - 3\right). \end{aligned}\right\} \quad \text{(I)}$$

Mahâvîra has observed that "when the areas are equal" we are to assume[1]

$$y = s\frac{m^2}{n^2},$$

$$x = s\frac{m^2}{n^2} - 1.$$

[1] Bibhutibhusan Datta has shown that this restriction is *not necessary*. In fact, starting with the assumption

$$\left.\begin{aligned} y &= s\frac{m^2q^2}{n^2p^2}, \\ x &= s\frac{m^2q^2}{n^2p^2} - 1; \end{aligned}\right\} m \geqslant n,\ q \geqslant p,$$

and proceeding in the same way as above, he has obtained another solution of (A) in the form

$$\left.\begin{aligned} y &= 2\frac{m^2q}{n^2p}, & v &= \frac{m}{n}, \\ x &= 2\frac{m^2q}{n^2p} - 1, & u &= \frac{m}{q}\left(4\frac{m^2q}{n^2p} - 2\right). \end{aligned}\right\} \quad \text{(II)}$$

Datta finds two general solutions of (A), *viz.*

$$\left.\begin{aligned} y &= \frac{rm^2q^2}{n^2p^2} + t, \\ x &= \frac{rm^2q^2}{tn^2p^2}\left(\frac{rm^2q^2}{n^2p^2} - \frac{rq}{p} + t\right), \\ v &= \frac{rmq}{np}, \\ u &= \frac{m}{nt}\left(\frac{rm^2q^2}{n^2p^2} + t\right)\left(\frac{rm^2q^2}{n^2p^2} - \frac{rq}{p} + t\right); \end{aligned}\right\} \quad \text{(III)}$$

Isosceles Triangles with Integral sides. Brahmagupta says :

"The sum of the squares of two unequal numbers is the side ; their product multiplied by two is the altitude, and twice the difference of the squares of those two unequal numbers is the base of an isosceles triangle."[1]

Mahâvîra gives the following rule for obtaining an isosceles triangle from a single generated rectangle:

"In the isosceles triangle (required), the two diagonals (of a generated rectangle[2]) are the two sides, twice its side is the base, the upright is the altitude, and the area (of the generated rectangle) is the area."[3]

Thus if m, n be two integers such that $m \neq n$, the sides of all rational isosceles triangles with integral sides are :

$$(i)\quad m^2+n^2,\ m^2+n^2,\ 2(m^2-n^2);$$

or

$$(ii)\quad m^2+n^2,\ m^2+n^2,\ 4mn.$$

and

$$\left.\begin{aligned} y &= \frac{rm^2q^2}{n^2p^2} - t, & v &= \frac{m}{n}\left(\frac{rq}{p} - t\right), \\ x &= \frac{rm^2q^2}{n^2p^2}\left(\frac{rq}{tp} - 1\right), & u &= \frac{rmq}{np}\left(\frac{rm^2q^2}{tn^2p^2} - 1\right) \end{aligned}\right\} \quad \text{(IV)}$$

where $m \geqslant n$, $q \geqslant p$ and r, t are any two integers.

See Datta, "On Mahâvîra's solution of rational triangles and quadrilaterals," *BCMS*, XX, 1928-9, pp. 267-294; particularly p. 285.

[1] *BrSpSi*, xii. 33.

[2] A rectangle generated from the numbers m and n has its sides equal to $m^2 - n^2$ and $2mn$ and its diagonal equal to $m^2 + n^2$. *Cf.* pp. 208-9.

[3] *GSS*, vii. 108½.

The altitude of the former is $2mn$ and of the latter $m^2 - n^2$ and the area in either case is the same, that is, $2mn(m^2 - n^2)$.

Juxtaposition of Right Triangles. It will be noticed that the device employed by Brahmagupta and Mahâvîra to find the above solutions is to juxtapose two rational right triangles—equal in this case—so as to have a common leg. It is indeed a very powerful device. For, every rational triangle or quadrilateral may be formed by the juxtaposition of two or four rational right triangles. So, in order to construct such rational figures, it suffices to know only the complete solution of $x^2 + y^2 = z^2$ in integers. The beginning of this principle is found as early as the *Baudhâyana Śulba*[1] (800 B.C.) wherein is described the formation of a kind of brick, called *ubhayî* (born of two), by the juxtaposition of the eighths of two suitable rectangular bricks of the same breadth (and thickness) but of different lengths.

Isosceles Triangles with a Given Altitude. Brahmagupta gives a rule to find all rational isosceles triangles having the same altitude. He says :

"The (given) altitude is the producer (*karaṇî*). Its square divided by an optional number is increased and diminished by that optional number. The smaller is the base and half the greater is the side."[2]

That is to say, the sides and bases of rational isosceles triangles having the same altitude a are respectively,

$$\tfrac{1}{2}\left(\frac{a^2}{m} + m\right),\ \tfrac{1}{2}\left(\frac{a^2}{m} + m\right) \text{ and } \left(\frac{a^2 - m}{m}\right),$$

where m is any rational number.

[1] *BŚl*, iii. 122 ; Compare Datta, *Śulba*, p. 45, where necessary figures are given.

[2] *BrSpSi*, xviii. 37.

In particular, let the given altitude be 8. Then taking $m = 4$ Pṛthûdakasvâmî (860) obtains the rational isosceles triangle (10, 10, 12).

Pairs of Rational Isosceles Triangles. Mahâvîra gives the following rule for finding two isosceles triangles whose perimeters, as also their areas, are related in given proportions :

"Multiply the square of the ratio-numbers of the perimeters by the ratio-numbers of the areas mutually and then divide the larger product by the smaller. Multiply the quotient by 6 and 2 (severally) and then diminish the smaller by unity : again (find severally) the difference between the results, and twice the smaller one : these are the two sets of elements for the figures to be generated. From them the sides, etc., can be obtained in the way described before."[1]

If (s_1, s_2) and $(\triangle_1, \triangle_2)$ denote the perimeters and areas of two rational isosceles triangles, such that

$$s_1 : s_2 = m : n, \qquad \triangle_1 : \triangle_2 = p : q, \qquad (1)$$

where the ratio-numbers m, n, p, q are known integers, then the triangles will be obtained, says Mahâvîra, from the rectangles *generated* from

$$\left(6\frac{n^2p}{m^2q},\ 2\frac{n^2p}{m^2q} - 1\right) \text{ and } \left(4\frac{n^2p}{m^2q} + 1,\ 4\frac{n^2p}{m^2q} - 2\right),$$

where $n^2p > m^2q$, when the dimensions of the first are multiplied by m and those of the second by n.

The dimensions of the isosceles triangle formed from the first set of *bîja* are :

$$\text{side} = m\left\{\left(6\frac{n^2p}{m^2q}\right)^2 + \left(2\frac{n^2p}{m^2q} - 1\right)^2\right\},$$

[1] *GSS*, vii. 137.

$$\text{base} \quad = 24m \frac{n^2p}{m^2q}\left(2\frac{n^2p}{m^2q} - 1\right),$$

$$\text{altitude} = m\left\{\left(6\frac{n^2p}{m^2q}\right)^2 - \left(2\frac{n^2p}{m^2q} - 1\right)^2\right\};$$

and from the second set

$$\text{side} \quad = n\left\{\left(4\frac{n^2p}{m^2q} + 1\right)^2 + \left(4\frac{n^2p}{m^2q} - 2\right)^2\right\},$$

$$\text{base} \quad = 4n\left(4\frac{n^2p}{m^2q} + 1\right)\left(4\frac{n^2p}{m^2q} - 2\right),$$

$$\text{altitude} = n\left\{\left(4\frac{n^2p}{m^2q} + 1\right)^2 - \left(4\frac{n^2p}{m^2q} - 2\right)^2\right\}.$$

It can be easily verified that the perimeters and areas of the isosceles triangles thus obtained satisfy the conditions (1).

In particular, putting $m = n = p = q = 1$, we have two isosceles triangles of sides, bases and altitudes (29, 40, 21) and (37, 24, 35) which have equal perimeters (98) and equal areas (420). This particular case was treated by Frans van Schooten the Younger (1657), J. H. Rahn (1697) and others.[1]

It is evident that multiplying the above values by m^4q^2 we get pairs of isosceles triangles whose dimensions are integral.

Rational Scalene Triangles. Brahmagupta says :

"The square of an optional number is divided twice by two arbitrary numbers; the moieties of the sums of the quotients and (respective) optional numbers are the sides of a scalene triangle; the sum of the moieties of the differences is the base."[2]

[1] Dickson, *Numbers*, II, p. 201.

[2] *BrSpSi*, xii. 34.

That is to say, the sides of a rational scalene triangle are

$$\tfrac{1}{2}\left(\frac{m^2}{p}+p\right),\ \tfrac{1}{2}\left(\frac{m^2}{q}+q\right),\ \tfrac{1}{2}\left(\frac{m^2}{p}-p\right)+\tfrac{1}{2}\left(\frac{m^2}{q}-q\right)$$

where m, p, q are any rational numbers. The altitude (m), area and segments of the base of this triangle are all rational.

Mahâvîra gives the rule :

"Half the base of a derived rectangle is divided by any optional number. With this divisor and the quotient is obtained another rectangle. The sum of the uprights (of these two rectangles) will be the base of the scalene triangle, the two diagonals, its sides and the base (of either rectangle) its altitude."[1]

If m, n be any two rational numbers, the rational rectangle ($AB'BH$)

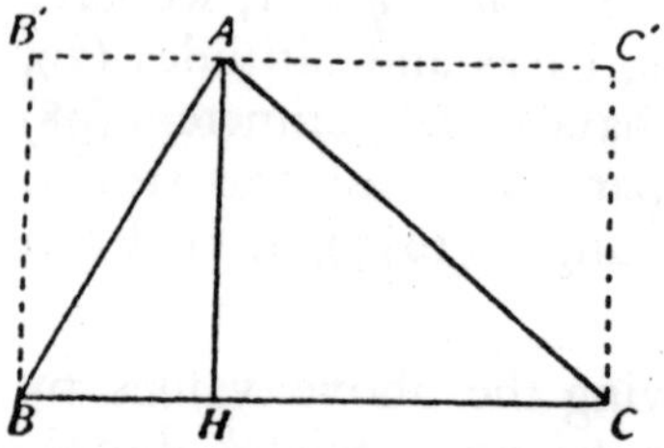

Fig. 3

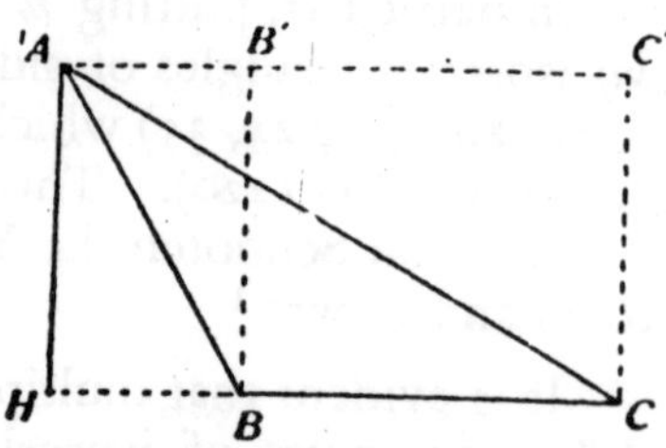

Fig. 4

formed from them is

$$m^2-n^2,\ 2mn,\ m^2+n^2.$$

If p, q be any two rational factors of mn, that is, if $mn=pq$, the second rectangle ($AC'CH$) is

$$p^2-q^2,\ 2pq,\ p^2+q^2.$$

[1] *GSS*, vii. 110½.

Now, juxtaposing these two rectangles so that they do not overlap (Fig. 3), the sides of the rational scalene triangle are obtained as

$$p^2 + q^2,\ m^2 + n^2,\ \{(p^2 - q^2) + (m^2 - n^2)\},$$

where $mn = pq$. Evidently the two rectangles can be juxtaposed so as to overlap (Fig. 4). So the general solution will be

$$p^2 + q^2,\ m^2 + n^2,\ \{(p^2 - q^2) \pm (m^2 - n^2)\}.$$

The altitude of the rational scalene triangle thus obtained is $2mn$ or $2pq$, its area $pq(p^2 - q^2) \pm mn(m^2 - n^2)$ and the segments of the base are $p^2 - q^2$ and $m^2 - n^2$.

In particular, putting $m = 12$, $p = 6$, $q = 8$ in Brahmagupta's general solution, Pṛthûdakasvâmî derives a scalene triangle of which the sides (13, 15), base (14), altitude (12), area (84) and the segments of the base (5, 9) are all integral numbers.

In order to get the above solutions of the rational scalene triangle the method employed was, it will be noticed, the juxtaposition of two rational right triangles so as to have a common leg. In Europe, it is found to have been employed first by Bachet (1621). The credit for the discovery of this method of finding rational scalene triangles should rightly go to Brahmagupta (628), but not to Bachet as is supposed by Dickson.[1]

Triangles having a Given Area. Mahâvîra proposes to find all triangles having the same given area A. His rules are:

"Divide the square of four times the given area by three; The quotient is the square of the square of a side of the equilateral triangle."[2]

[1] Dickson, *Numbers*, II, p. 192.
[2] *GSS*, vii. 154½.

"Divide the given area by an optional number; the square-root of the sum of the squares of the quotient and the optional number is a side of the isosceles triangle formed. Twice the optional number is its base and the area divided by the optional number is the altitude."[1]

"The cube of the square-root of the sum of eight times the given area and the square of an optional number is divided by the product of the optional number and that square-root; the quotient is diminished by half the optional number which is the base (of the required triangle). The *saṅkramaṇa* between this remainder and the quotient of the square of the optional number divided by twice that square-root will give the two sides."[2]

The last rule has been re-stated differently.[3]

21. RATIONAL QUADRILATERALS

Rational Isosceles Trapeziums. Brahmagupta has shown how to obtain an isosceles trapezium whose sides, diagonals, altitude, segments and area are all rational numbers. He says:

"The diagonals of the rectangle (generated) are the flank sides of an isosceles trapezium; the square of its side is divided by an optional number and then lessened by that optional number and divided by two; (the result) increased by the upright is the base and lessened by it is the face."[4]

That is to say, we shall have (Fig. 5)

$$CD = \tfrac{1}{2}\left(\frac{4m^2n^2}{p} - p\right) + (m^2 - n^2),$$

[1] *GSS*, vii. 156½.
[2] *GSS*, vii. 158½.
[3] *GSS*, vii. 160½-161½.
[4] *BrSpSi*, xii. 36.

$$AB = \tfrac{1}{2}\Big(\frac{4m^2n^2}{p} - p\Big) - (m^2 - n^2),$$

$$AD = BC = m^2 + n^2\,;$$

also

$$DH = m^2 - n^2,$$

$$HC = \tfrac{1}{2}\Big(\frac{4m^2n^2}{p} - p\Big),$$

$$AC = BD = \tfrac{1}{2}\Big(\frac{4m^2n^2}{p} + p\Big),$$

$$AH = 2mn,$$

area $$ABCD = mn\Big(\frac{4m^2n^2}{p} - p\Big).$$

By choosing the values of m, n and p suitably, the values of all the dimensions of the isosceles trapezium can be made integral. Thus, starting with the rectangle (5, 12, 13) and taking $p = 6$, Pṛthûdakasvâmî finds, by way of illustration, the isosceles trapezium whose flank sides = 13, base = 14, and face = 4. Its altitude (12), segments of base (5, 9), diagonals (15) and area (108) are also integers.

Mahâvîra writes:

"For an isosceles trapezium the sum of the perpendicular of the first generated rectangle and the perpendicular of the second rectangle which is generated from any (rational) divisor of half the base of the first and the quotient, will be the base; their difference will be the face; the smaller of the diagonals (of the generated rectangles) will be the flank side; the smaller perpendicular will be the segment; the greater diagonal will be the diagonal (of the isosceles trapezium); the greater area will be the area and the base (of either rectangle) will be the altitude."[1]

[1] *GSS*, vii. 99½.

The first rectangle ($AA'DH$) generated from m, n is

$$m^2 - n^2, \ 2mn, \ m^2 + n^2.$$

If p, q be any two rational factors of half the base of this rectangle, that is, if $pq = mn$, the second rectangle ($AB'CH$) from these factors will be

$$p^2 - q^2, \ 2pq, \ p^2 + q^2.$$

By judiciously juxtaposing these two rectangles, we shall obtain an isosceles trapezium of the type required ($ABCD$):

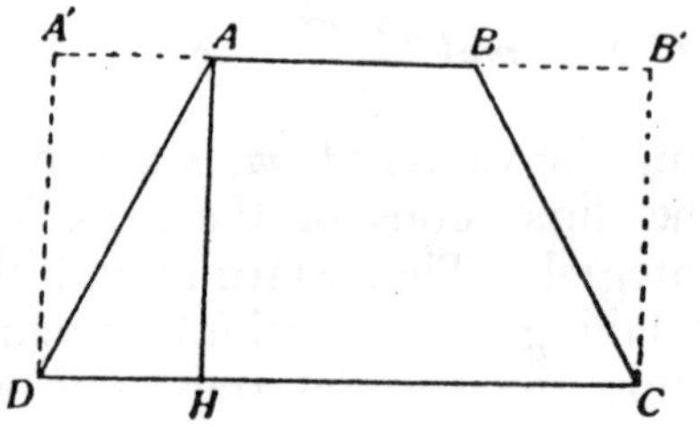

Fig. 5

$$\begin{aligned}
CD &= (p^2 - q^2) + (m^2 - n^2),\\
AB &= (p^2 - q^2) - (m^2 - n^2),\\
AD = BC &= m^2 + n^2, \text{ if } m^2 + n^2 < p^2 + q^2,\\
DH &= m^2 - n^2, \quad \text{ if } m^2 - n^2 < p^2 - q^2,\\
AC = BD &= p^2 + q^2, \text{ if } p^2 + q^2 > m^2 + n^2,\\
AH &= 2mn = 2pq,\\
\text{area } ABCD &= 2pq\,(p^2 - q^2),\\
&\qquad \text{if } 2pq\,(p^2 - q^2) > 2mn(m^2 - n^2).
\end{aligned}$$

The necessity of the conditions $m^2 + n^2 < p^2 + q^2$, $m^2 - n^2 < p^2 - q^2$, etc., will be at once realised from a glance at Figs. 5 and 6. The above specifications of the dimensions of a rational isosceles trapezium will give Fig. 5. But when the conditions are reversed so that

$m^2 + n^2 > p^2 + q^2$, $m^2 - n^2 > p^2 - q^2$, $2pq\,(p^2 - q^2) < 2mn(m^2 - n^2)$, the dimensions of the isosceles trapezium (Fig. 6) are :

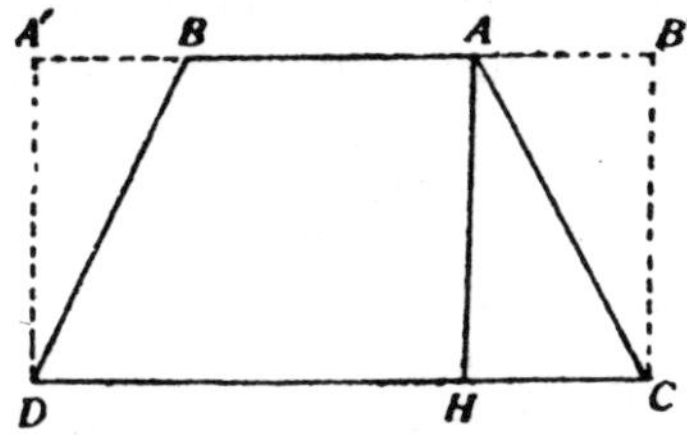

Fig. 6

$$
\begin{aligned}
CD &= (m^2 - n^2) + (p^2 - q^2),\\
AB &= (m^2 - n^2) - (p^2 - q^2),\\
AC &= BD = p^2 + q^2,\\
DH &= m^2 - n^2,\\
AD &= BC = m^2 + n^2,\\
AH &= 2mn = 2pq,\\
\text{area } ABCD &= 2mn(m^2 - n^2).
\end{aligned}
$$

Pairs of Isosceles Trapeziums. Mahâvîra gives the following rule for finding the face, base and equal sides of an isosceles trapezium having an area and altitude exactly equal to those of another isosceles trapezium whose dimensions are known :

"On performing the *viṣama-saṅkramaṇa* between the square of the perpendicular (of the known isosceles trapezium) and an optional number, the greater result will be the equal sides of the (required) isosceles trapezium; half the sum and difference of the smaller result and the moieties of the face and base (of the known figure) will be the base and face (respectively of the required figure)."[1]

[1] *GSS*, vii. 173½.

Let a, b, c, h, denote respectively the face, base, equal sides and altitude of the known isosceles trapezium and let a', b', c', h', denote the corresponding quantities of the required isosceles trapezium. Then, since the two trapeziums are equal in area and altitude, we must have

$$h' = h,$$

$$b' + a' = b + a, \tag{1}$$

and $$c'^2 - \left(\frac{b' - a'}{2}\right)^2 = h^2,$$

or $$\{c' + (b' - a')/2\}\{c' - (b' - a')/2\} = h^2,$$

whence $$c' - (b' - a')/2 = r,$$

and $$c' + (b' - a')/2 = h^2/r,$$

r being any rational number. Then

$$c' = \tfrac{1}{2}(h^2/r + r), \tag{2}$$

$$b' - a' = (h^2/r - r). \tag{3}$$

From (1) and (2), we get

$$b' = (b + a)/2 + (h^2/r - r)/2, \tag{4}$$

$$a' = (b + a)/2 - (h^2/r - r)/2. \tag{5}$$

If $a = 4$, $b = 14$, $c = 13$, $h = 12$, taking $r = 10$, we shall have[1] $a' = 34/5$, $b' = 56/5$, $c' = 61/5$.

It has been stated above that, if m, n, p, q are rational numbers such that $m^2 \pm n^2 < p^2 \pm q^2$, we must have

$$a = (p^2 - q^2) - (m^2 - n^2),$$

$$b = (p^2 - q^2) + (m^2 - n^2),$$

$$c = m^2 + n^2,$$

$$h = 2mn = 2pq.$$

[1] *GSS*, vii. 174½.

Substituting these values in (2), (4), (5) we get the dimensions of the equivalent isosceles trapezium as

$$a' = (p^2 - q^2) - (4p^2q^2/r - r)/2,$$
$$b' = (p^2 - q^2) + (4p^2q^2/r - r)/2,$$
$$c' = (4p^2q^2/r + r)/2.$$

If $m^2 \pm n^2 > p^2 \pm q^2$, the sides of the pair of isosceles trapeziums equal in area and altitude will be

$$a = (m^2 - n^2) - (p^2 - q^2),$$
$$b = (m^2 - n^2) + (p^2 - q^2),$$
$$c = p^2 + q^2;$$
$$a' = (m^2 - n^2) - (4m^2n^2/r - r)/2,$$
$$b' = (m^2 - n^2) + (4m^2n^2/r - r)/2,$$
$$c' = (4m^2n^2/r + r)/2.$$

These two isosceles trapeziums will also have equal diagonals.

Rational Trapeziums with Three Equal Sides. This problem is nearly the same as that of the rational isosceles trapezium with this difference that in this case one of the parallel sides also is equal to the slant sides. Brahmagupta states the following solution of the problem:

"The square of the diagonal (of a *generated* rectangle) gives three equal sides; the fourth (is obtained) by subtracting the square of the upright from thrice the square of the side (of that rectangle). If greater, it is the base; if less, it is the face."[1]

The rectangle generated from m, n is given by

$$m^2 - n^2, \ 2mn, \ m^2 + n^2.$$

[1] *BrSpSi*, xii. 37.

If $ABCD$ be a rational trapezium whose sides AB, BC, AD are equal, then

$$AB = BC = AD = (m^2 + n^2)^2,$$

$$CD = 3(2mn)^2 - (m^2 - n^2)^2 = 14m^2n^2 - m^4 - n^4,$$

or $$CD = 3(m^2 - n^2)^2 - (2mn)^2 = 3m^4 + 3n^4 - 10m^2n^2.$$

In particular, putting $m = 2$, $n = 1$, Pṛthûdakasvâmî deduces two rational trapeziums with three equal sides, *viz.*, (25, 25, 25, 39) and (25, 25, 25, 11).

The first solution is also given by Mahâvîra who indicates the method for obtaining it. He says:

"For a trapezium with three equal sides (proceed) as in the case of the isosceles trapezium with (the rectangle formed from) the quotient of the area of a generated rectangle divided by the square-root of its side multiplied by the difference of its elements and divisor; and that (formed) from the side and upright."[1]

That is to say, from any rectangle $(m^2 - n^2, 2mn, m^2 + n^2)$, calculate

$$\frac{2mn(m^2 - n^2)}{\sqrt{2mn(m - n)}} = \sqrt{2mn}\,(m + n).$$

Then from $\sqrt{2mn(m - n)}$, $\sqrt{2mn}(m + n)$ form the rectangle

$$8m^2n^2,\ 4mn(m^2-n^2),\ 4mn(m^2+n^2). \tag{1}$$

Again from $2mn$, $m^2 - n^2$ form another rectangle

$$6m^2n^2 - m^4 - n^4,\ 4mn\,(m^2 - n^2),\ (m^2 + n^2)^2 \tag{2}$$

By the juxtaposition of the rectangles (1) and (2) we get Brahmagupta's rational trapezium with three

[1] *GSS*, vii. 101½.

equal sides :

$$CD = 8mn^2 + (6m^2n^2 - m^4 - n^4) = 14m^2n^2 - m^4 - n^4,$$

$$AB = 8m^2n^2 - (6m^2n^2 - m^4 - n^4) = (m^2 + n^2)^2 = AD$$

$$= BC, \text{ if } m^2 + n^2 < 4mn.$$

The segment (CH), altitude (AH), diagonals (AC, BD) and area of this trapezium are also rational, for

$$CH = 6m^2n^2 - m^4 - n^4,$$

$$AH = 4mn(m^2 - n^2),$$

$$AC = BD = 4mn(m^2 + n^2),$$

area $$ABCD = 32m^3n^3(m^2 - n^2).$$

Rational Inscribed Quadrilaterals. Brahmagupta formulated a remarkable proposition: To find all quadrilaterals which will be inscribable within circles and whose sides, diagonals, perpendiculars, segments (of sides and diagonals by perpendiculars from vertices as also of diagonals by their intersection), areas, and also the diameters of the circumscribed circles will be expressible in integers. Such quadrilaterals are called Brahmagupta quadrilaterals.

The solution given by Brahmagupta is as follows :

"The uprights and bases of two right-angled triangles being reciprocally multiplied by the diagonals of the other will give the sides of a quadrilateral of unequal sides: (of these) the greatest is the base, the least is the face, and the other two sides are the two flanks."[1]

Taking Brahmagupta's integral solution, the sides of the two right triangles of reference are given by

$$m^2 - n^2, \ 2mn, \ m^2 + n^2;$$

$$p^2 - q^2, \ 2pq, \ p^2 + q^2;$$

[1] *BrSpSi*, xii. 38.

where m, n, p, q are integers. Then the sides of a Brahmagupta quadrilateral are

$$\left.\begin{array}{c}(m^2 - n^2)(p^2 + q^2), \quad (p^2 - q^2)(m^2 + n^2), \\ 2mn(p^2 + q^2), \quad 2pq(m^2 + n^2).\end{array}\right\} \quad (A)$$

Mahâvîra says :

"The base and the perpendicular (of the smaller and the larger derived rectangles of reference) multiplied reciprocally by the longer and the shorter diagonals and (each again) by the shorter diagonal will be the sides, the face and the base (of the required quadrilateral). The uprights and bases are reciprocally multiplied and then added together; again the product of the uprights is added to the product of the bases; these two sums multiplied by the shorter diagonal will be the diagonals. (These sums) when multiplied respectively by the base and perpendicular of the smaller figure of reference will be the altitudes; and they when multiplied respectively by the perpendicular and the base will be the segments of the base. Other segments will be the difference of these and the base. Half the product of the diagonals (of the required figure) will be the area."[1]

If the rectangle generated from m, n be smaller than that from p, q, then, according to Mahâvîra, we obtain the rational inscribed quadrilateral of which the sides are

$$(m^2 - n^2)(p^2 + q^2)(m^2 + n^2), \quad (p^2 - q^2)(m^2 + n^2)^2,$$
$$2mn(p^2 + q^2)(m^2 + n^2), \quad 2pq(m^2 + n^2)^2;$$

whose diagonals are

$$\{2pq(m^2 - n^2) + 2mn(p^2 - q^2)\}(m^2 + n^2),$$
$$\{(m^2 - n^2)(p^2 - q^2) + 4mnpq\}(m^2 + n^2);$$

[1] *GSS*, vii. 103½.

whose altitudes are

$$\{2pq(m^2 - n^2) + 2mn(p^2 - q^2)\}2mn,$$
$$\{(m^2 - n^2)(p^2 - q^2) + 4mnpq\}(m^2 - n^2);$$

whose segments are

$$\{2pq(m^2 - n^2) + 2mn(p^2 - q^2)\}(m^2 - n^2),$$
$$\{(m^2 - n^2)(p^2 - q^2) + 4mnpq\}2mn;$$

and whose area is

$$\tfrac{1}{2}\{2pq(m^2 - n^2) + 2mn(p^2 - q^2)\}\{(m^2 - n^2)(p^2 - q^2) + 4mnpq\}(m^2 + n^2)^2.$$

Śrîpati writes :

"Of the two right triangles the sides and uprights are reciprocally multiplied by the hypotenuses; of the products the greatest is the base, the smallest is the face and the rest are the two flank sides of a quadrilateral with unequal sides."[1]

Bhâskara II gives the rule :

"The sides and uprights of two optional right triangles being multiplied by their reciprocal hypotenuses become the sides : in this way has been derived a quadrilateral of unequal sides. There the two diagonals can be obtained from those two triangles. The product of the uprights, added with the product of the sides, gives one diagonal; the sum of the reciprocal products of the uprights and sides is the other."[2]

Bhâskara II[3] illustrates by taking the right triangles (3, 4, 5) and (5, 12, 13) so that the resulting cyclic quadrilateral is (25, 39, 60, 52). The same example was

[1] *SiŚe*, xiii. 42.
[2] *L*, p. 51.
[3] *L*, p. 52.

given before by Mahâvîra[1] and Pṛthûdakasvâmî.[2] This cyclic quadrilateral also appears in the *Triśatikâ* of Srîdhara[3] and in the commentary of the *Âryabhaṭîya* by Bhâskara I (522). The diagonals of this quadrilateral are, states Bhâskara II, 56 (= 3.12 + 4.5) and 63 (= 4.12 + 3.5) (Fig. 7). He then observes:

"If the figure be formed by changing the arrangement of the face and flank then the second diagonal will be equal to the product of the hypotenuses of the two right triangles (of reference), *i.e.*, 65." (Fig. 8).

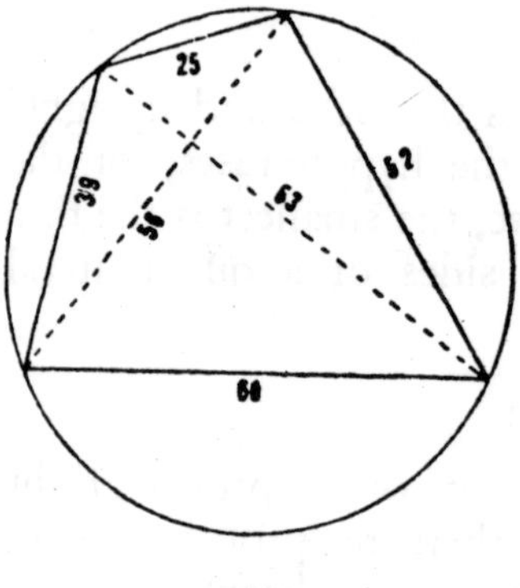

Fig. 7

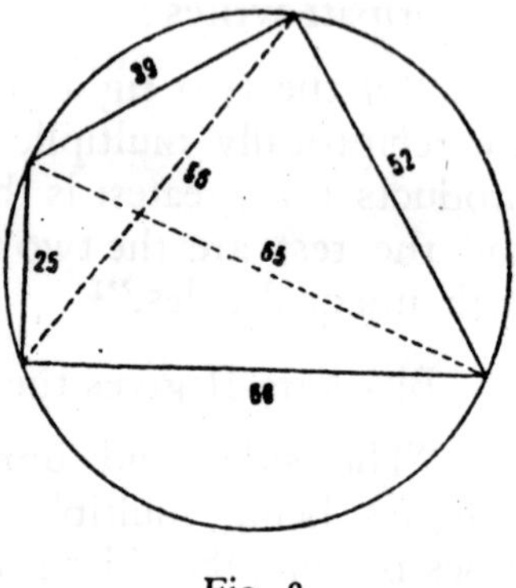

Fig. 8

By taking the right triangles (3, 4, 5) and (15, 8, 17) Bhâskara II gets another cyclic quadrilateral (68, 51, 40, 75), whose diagonals are (77, 85), altitude is 308/5, segments are 144/5 and 231/5, and area is 3234.[4] (Fig. 9). With the sequence of the sides (68, 40, 51, 75) the

[1] *GSS*, vii. 104½.
[2] *BrSpSi*, xii. 38 (*Com.*).
[3] *Triś*, Ex. 80.
[4] *L*, pp. 46ff.

diagonals are (77, 84) (Fig. 10), and with (68, 40, 75, 51) they are (84, 85). (Fig. 11).

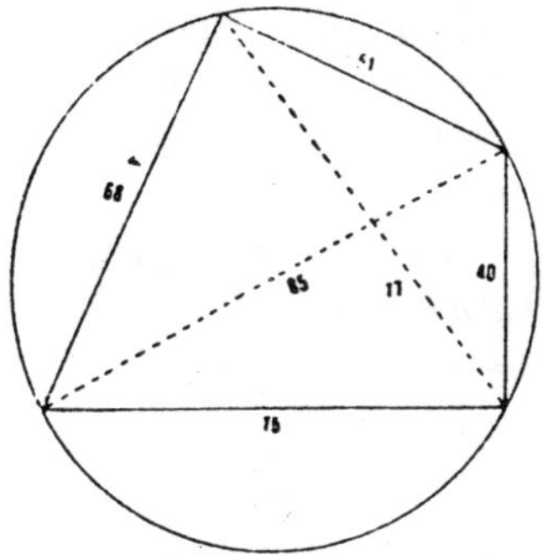

Fig. 9

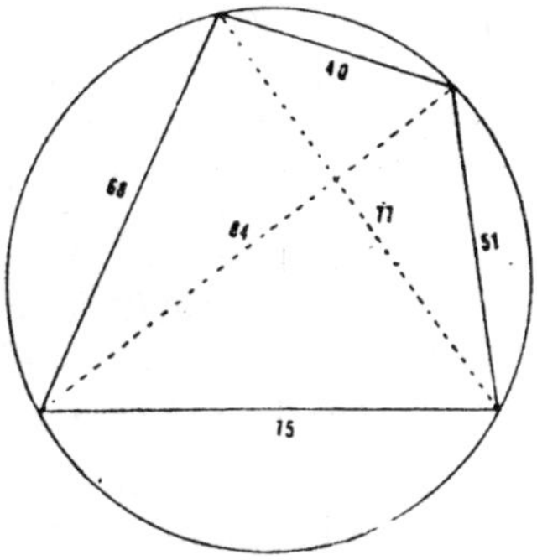

Fig. 10

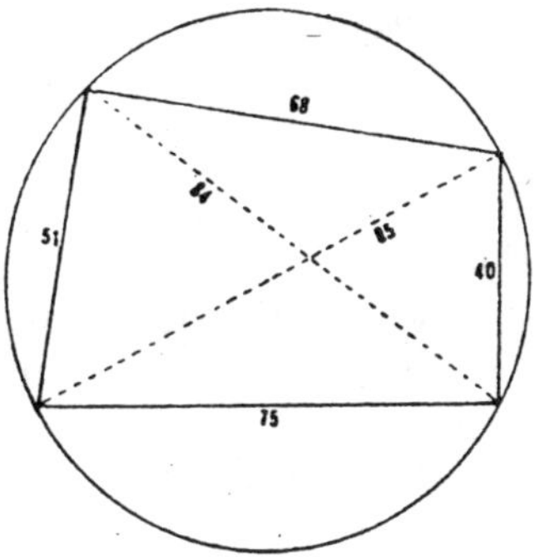

Fig. 11

The deep significance of Brahmagupta's results has been demonstrated by Chasles[1] and Kummer.[2]

[1] M. Chasles, *Aperçu historique sur l'origine et development des méthodes en géométrie*, Paris, 1875, pp. 436ff.

[2] E. E. Kummer, "Über die Vierecke, deren Seiten und Diogonalen rational sind," *Journ. für Math.*, XXXVII, 1848, pp. 1-20.

In fact, according to the sequence in which the quantities (A) are taken, there will be two varieties of Brahmagupta quadrilaterals having them as their sides: (1) one in which the two diagonals intersect at right angles and (2) the other in which the diagonals intersect obliquely. The arrangement (A) gives a quadrilateral of the first variety. For the oblique variety, the sides are in the following order:

$$\left.\begin{array}{ll}(p^2-q^2)(m^2+n^2), & (m^2-n^2)(p^2+q^2),\\ 2mn(p^2+q^2), & 2pq(m^2+n^2);\end{array}\right\} \quad (B)$$

or

$$\left.\begin{array}{ll}(p^2-q^2)(m^2+n^2), & 2mn(p^2+q^2),\\ (m^2-n^2)(p^2+q^2), & 2pq(m^2+n^2).\end{array}\right\} \quad (C)$$

Bhâskara II points out that the diagonals of the Brahmagupta quadrilateral are in the (A) variety,

$$2pq(m^2-n^2)+2mn(p^2-q^2), \quad 4mnpq+(p^2-q^2)(m^2-n^2);$$

in (B),

$$2pq(m^2-n^2)+2mn(p^2-q^2), \quad (p^2+q^2)(m^2+n^2);$$

and in (C),

$$4mnpq+(p^2-q^2)(m^2-n^2), \quad (p^2+q^2)(m^2+n^2).$$

The diameter of the circumscribed circle in every case is $(p^2+q^2)(m^2+n^2)$.

Gaṇeśa (1545) shows that the quadrilateral is formed by the juxtaposition of four right triangles obtained by multiplying the sides of each of two rational right triangles by the upright and base of the other.[1] He writes:

"A quadrilateral is divided into four triangles by its intersecting diagonals. So by the juxtaposition of four triangles a quadrilateral will be formed. For that purpose the four triangles are assumed in this manner: Take two right triangles formed in the way indicated

[1] Compare also L. E. Dickson, "Rational Triangles and Quadrilaterals," *Amer. Math. Mon.*, XXVIII, 1921, pp. 244-250.

before. If the upright, base and hypotenuse of a rational right triangle be multiplied by any arbitrary rational number, there will be produced another right triangle with rational sides. Hence on multiplying the sides of each of the two right triangles by an optional number equal to the base of the other and again by an optional nnmber equal to the upright of the other, four right triangles will be obtained by the judicious juxtaposition of which the quadrilateral will be formed."

He then shows with the help of specific examples (see Figs. 12, 13 & 14) that we can obtain in this way

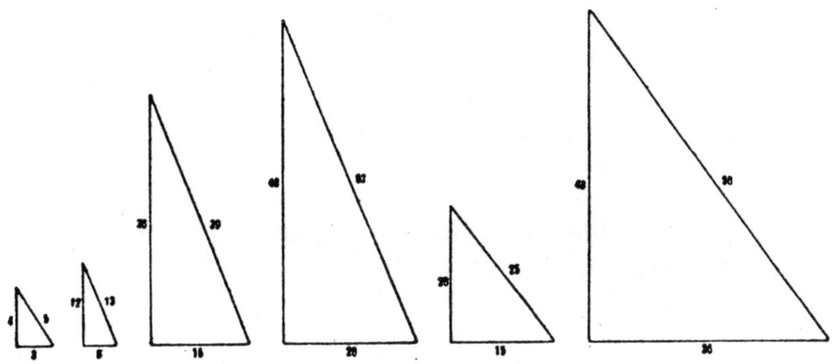

Fig. 12

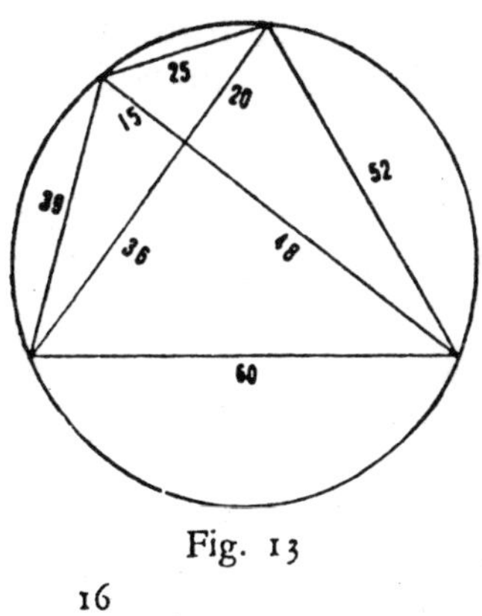

Fig. 13

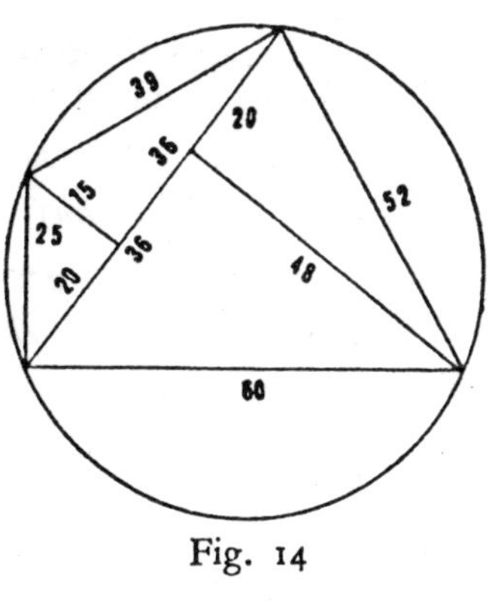

Fig. 14

from the same set of two rational right triangles two varieties of rational convex quadrilaterals: One in which the diagonals intersect each other perpendicularly; and the other in which they do so obliquely.

Inscribed Quadrilaterals having a Given Area. Mahâvîra proposes to find all rational inscribed rectangles having the same given area (A, say). He says:

"The square-root of the exact area is a side of the square. The quotient of the area by an optional number and that optional number will be the base and upright of the rectangle."[1]

For finding all inscribed rational isosceles trapeziums having the same area A his rule is:

"The given area multiplied by the square of an optional number is diminished by the area of a generated rectangle and then divided by the base of that rectangle; the quotient divided by the optional number is the face; the quotient added with twice the upright and divided by the optional number gives the base; the base (of the generated rectangle) divided by the optional number is the altitude; and the diagonal divided by the optional number gives the two flank sides."[2]

That is to say, if $m^2 - n^2$, $2mn$, $m^2 + n^2$ be the upright, base and diagonal of a rectangle formed from m, n, the dimensions of the isosceles trapezium will be

$$\text{face} = \frac{s^2A - 2mn(m^2 - n^2)}{2mns},$$

$$\text{base} = \frac{1}{s}\left\{\frac{s^2A - 2mn(m^2 - n^2)}{2mn} + 2(m^2 - n^2)\right\}$$

$$= \frac{s^2A + 2mn(m^2 - n^2)}{2mns},$$

[1] *GSS*, vii. 146. [2] *GSS*, vii. 148.

$$\text{altitude} = \frac{2mn}{s},$$

$$\text{side} \quad = \frac{m^2+n^2}{s};$$

where s is an arbitrary rational number chosen such that $s^2A > 2mn(m^2-n^2)$.

For the construction of an inscribed trapezium of three equal sides Mahâvîra gives the following rule :

"The square of the given area is divided by the cube of an optional number and then increased by that optional number ; half the result gives the (equal) sides of a generated trapezium of three equal sides (having the given area) ; twice the optional number diminished by the side is the base ; and the given area divided by the optional number is the altitude."[1]

In other words, the dimensions of an inscribed trapezium of three equal sides having a given area A will be

$$\text{side} \quad = \tfrac{1}{2}\left(\frac{A^2}{s^3} + s\right),$$

$$\text{base} \quad = 2s - \tfrac{1}{2}\left(\frac{A^2}{s^3} + s\right),$$

$$\text{altitude} = \frac{A}{s}.$$

To find inscribed quadrilaterals having a given area Mahâvîra gives the following rule :

"Break up the square of the given area into any four arbitrary factors. Half the sum of these factors is diminished by them (severally). The remainders are the sides of an (inscribed) quadrilateral with unequal sides."[2]

[1] *GSS*, vii. 150.

[2] *GSS*, vii. 152. This result follows from the fact that the area of a cyclic quadrilateral is $\sqrt{(s-a)(s-b)(s-c)(s-d)}$.

Triangles and Quadrilaterals having a Given Circum-Diameter. Mahâvîra proposes to find all rational triangles and quadrilaterals inscribable in a circle of given diameter. His solution is:

"Divide the given diameter of the circle by the calculated diameter (of the circle circumscribing any generated figure of the required kind). The sides of that generated figure multiplied by the quotient will be the sides of the required figure."[1]

In other words, we shall have to find first a rational triangle or cyclic quadrilateral; then calculate the diameter of its circum-circle and divide the given diameter by it. Dimensions of the optional figure multiplied by this quotient will give the dimensions of the required figure of the type.

It has been found before (p. 227) that the sides of a rational triangle are proportional to

$$m^2 + n^2,\ p^2 + q^2,\ (p^2 - q^2) \pm (m^2 - n^2)$$

and its altitude is proportional to $2mn$ (or $2pq$), m, n, p, q being any rational numbers such that $mn = pq$. The diameter of the circle circumscribed about this triangle is proportional to

$$\frac{(m^2 + n^2)(p^2 + q^2)}{2mn}.$$

Then the sides of a rational triangle inscribed in a circle of diameter D will be

$$\frac{2mnD}{p^2 + q^2},\ \frac{2mnD}{m^2 + n^2},\ 2mnD\frac{(p^2 - q^2) \pm (m^2 - n^2)}{(m^2 + n^2)(p^2 + q^2)};$$

and its altitude will be

$$\frac{D(2mn)^2}{(m^2 + n^2)(p^2 + q^2)}.$$

[1] *GSS*, vii. 221½.

The dimensions of a rational inscribed quadrilateral, as stated by Mahâvîra, have been noted before (p. 236). The diameter of its circum-circle is

$$(p^2 + q^2)(m^2 + n^2)^2.$$

Then, according to Mahâvîra, the sides of a rational quadrilateral inscribed in a circle of diameter D, are

$$D\left(\frac{2mn}{m^2 + n^2}\right),\ D\left(\frac{m^2 - n^2}{m^2 + n^2}\right),\ D\left(\frac{2pq}{p^2 + q^2}\right),\ D\left(\frac{p^2 - q^2}{p^2 + q^2}\right);$$

its diagonals are

$$\{2pq\,(m^2 - n^2) + 2mn\,(p^2 - q^2)\}\frac{D}{(p^2 + q^2)(m^2 + n^2)},$$

$$\{(m^2 - n^2)(p^2 - q^2) + 4mnpq\}\frac{D}{(p^2 + q^2)(m^2 + n^2)};$$

and its area is

$$\frac{D^2}{2(p^2 + q^2)(m^2 + n^2)}\{2pq(m^2 - n^2) + 2mn(p^2 - q^2)\}$$
$$\times\{(m^2 - n^2)(p^2 - q^2) + 4mnpq\};$$

so that the sides, diagonals and area are all rational.

22. SINGLE INDETERMINATE EQUATIONS OF HIGHER DEGREES

The Hindus do not seem to have paid much attention to the treatment of indeterminate equations of degrees higher than the second. Some interesting examples involving such equations are, however, found in the works of Mahâvîra (850), Bhâskara II (1150) and Nârâyaṇa (1350).

Mahâvîra's Rule. One problem of Mahâvîra is as follows :

Given the sum (s) of a series in A.P., to find its

first term (a), common difference (b) and the number of terms (n).

In other words, it is required to solve in rational numbers the equation

$$\{a + (\frac{n-1}{2})b\}n = s,$$

containing three unknowns a, b and n, and of the third degree. The following rule is given for its solution :

"Here divide the sum by an optional factor of it ; that divisor is the number of terms. Subtract from the quotient another optional number ; the subtrahend is the first term. The remainder divided by the half of the number of terms as diminished by unity is the common difference."[1]

Bhâskara's Method. Bhâskara II proposes the problems :

"Tell those four numbers which are unequal but have a common denominator, whose sum or the sum of whose cubes is equal to the sum of their squares."[2]

If x, y, z, w be the numbers, then

(1) $x + y + z + w = x^2 + y^2 + z^2 + w^2,$

(2) $x^3 + y^3 + z^3 + w^3 = x^2 + y^2 + z^2 + w^2.$

Let the numbers be x, $2x$, $3x$, $4x$, says Bhâskara II. That is, suppose $y = 2x$, $z = 3x$, $w = 4x$ in the above

[1] *GSS*, vii. 78.

There are also other problems where instead of s, the given quantity is $s + a$, $s + b$, $s + n$ or $s + a + b + n$. (*GSS*, ii. 83 ; *cf.* also vi. 80). For such problems also the method of solution is the same as before, *i.e.*, to assume suitable arbitrary values for two of the unknowns so that the indeterminate cubic equation is thereby reduced to a determinate linear equation in one unknown. (*GSS*, ii. 82 ; vi. 317).

[2] *BBi*, p. 55.

equations. Then by (1) we get

$$10x = 30x^2.$$

$$\therefore \quad x = \tfrac{1}{3}.$$

Hence $\qquad x, y, z, w = \frac{1}{3}, \frac{2}{3}, \frac{3}{3}, \frac{4}{3},$

is a solution of (1).

Again, with the same assumption, the equation (2) reduces to

$$100x^3 = 30x^2.$$

$$\therefore \quad x = \tfrac{3}{10}.$$

Hence $\qquad x, y, z, w = \frac{3}{10}, \frac{6}{10}, \frac{9}{10}, \frac{12}{10},$

is a solution of (2).

The following problem has been quoted by Bhâskara II from an ancient author :

"The square of the sum of two numbers added with the cube of their sum is equal to twice the sum of their cubes. Tell, O mathematician, (what are those two numbers)."[1]

If x, y be the numbers, then by the statement of the question

$$(x+y)^2 + (x+y)^3 = 2(x^3+y^3).$$

"Here, so that the operations may not become lengthy," says Bhâskara II, "assume the two numbers to be $u+v$ and $u-v$." So on putting

$$x = u+v, \ y = u-v,$$

the equation reduces to

$$4u^3 + 4u^2 = 12uv^2,$$

or $\qquad 4u^2 + 4u = 12v^2,$

or $\qquad (2u+1)^2 = 12v^2 + 1.$

[1] *BBi*, p. 101.

Solving this equation by the method of the Square-nature we get values of u, v. Whence the values of (x, y) are found to be (5, 1), (76, 20), etc.

Nârâyaṇa's Rule. Nârâyaṇa gives the rule:

"Divide the sum of the squares, the square of the sum and the product of any two optional numbers by the sum of their cubes and the cube of their sum, and then multiply by the two numbers (severally). (The results) will be the two numbers, the sum of whose cubes and the cube of whose sum will be equal to the sum of their squares, the square of the sum and the product of them."[1]

That is to say, the solution of the equations

(1) $x^3 + y^3 = x^2 + y^2$, (4) $(x + y)^3 = x^2 + y^2$,

(2) $x^3 + y^3 = (x + y)^2$, (5) $(x + y)^3 = (x + y)^2$,

(3) $x^3 + y^3 = xy$, (6) $(x + y)^3 = xy$,

are respectively

$$(1.1) \quad \begin{cases} x = \dfrac{(m^2 + n^2)m}{m^3 + n^3}, \\[2ex] y = \dfrac{(m^2 + n^2)n}{m^3 + n^3}; \end{cases} \qquad (4.1) \quad \begin{cases} x = \dfrac{(m^2 + n^2)m}{(m + n)^3}, \\[2ex] y = \dfrac{(m^2 + n^2)n}{(m + n)^3}; \end{cases}$$

$$(2.1) \quad \begin{cases} x = \dfrac{(m + n)^2 m}{m^3 + n^3}, \\[2ex] y = \dfrac{(m + n)^2 n}{m^3 + n^3}; \end{cases} \qquad (5.1) \quad \begin{cases} x = \dfrac{(m + n)^2 m}{(m + n)^3}, \\[2ex] y = \dfrac{(m + n)^2 n}{(m + n)^3}; \end{cases}$$

$$(3.1) \quad \begin{cases} x = \dfrac{m^2 n}{m^3 + n^3}, \\[2ex] y = \dfrac{m n^2}{m^3 + n^3}; \end{cases} \qquad (6.1) \quad \begin{cases} x = \dfrac{m^2 n}{(m + n)^3}, \\[2ex] y = \dfrac{m n^2}{(m + n)^3}; \end{cases}$$

[1] *GK*, i. 58.

where m, n are rational numbers.

It will be noticed that the equation (2) can be reduced, by dividing out by $x+y$, to

$$x^2 - xy + y^2 = x + y;$$

and similarly (5) can be reduced to

$$x + y = 1.$$

With $m = 1$, $n = 2$ Nârâyaṇa gives the following sets of particular values:

(1.2) $x, y = \frac{5}{9}, \frac{10}{9}$ (4.2) $x, y = \frac{5}{27}, \frac{10}{27}$

(2.2) $x, y = 1, 2$ (5.2) $x, y = \frac{1}{3}, \frac{2}{3}$,

(3.2) $x, y = \frac{2}{9}, \frac{4}{9}$ (6.2) $x, y = \frac{2}{27}, \frac{4}{27}$.

He then observes: "In this way one can find by his own intelligence two numbers for the case of *difference*, etc."

Form $ax^{2n+2} + bx^{2n} = y^2$. For the solution of an equation of the form

$$ax^{2n+2} + bx^{2n} = y^2,$$

where n is an integer, Bhâskara II gives the following rule:

"Removing a square factor from the second side, if possible, the two roots should be investigated in this case. Then multiply the greater root by the lesser. Or, if a biquadratic factor has been removed, the greater root should be multiplied by the square of the lesser root. The rest of the operations will then be as before."[1]

Suppose $ax^2 + b = z^2$; then the equation becomes

$$y^2 = x^{2n}z^2.$$

$$\therefore \quad y = x^n z.$$

The method of solving $ax^2 + b = z^2$ in positive integers has been described before.

[1] *BBi*, p. 102.

Two examples of equations of this form occur in the *Bîjagaṇita* of Bhâskara II :[1]

(1) $5x^4 - 100x^2 = y^2$,

(2) $8x^6 + 49x^4 = y^2$.

It may be noted that the second equation appears in the course of solving another problem.

Equation $ax^4 + bx^2 + c = u^3$. One very special case of this form arises in the course of solving another problem. It is[2]

$$(a + x^2)^2 + a^2 = u^3,$$

or

$$x^4 + 2ax^2 + 2a^2 = u^3.$$

Let $u = x^2$, supposes Bhâskara II, so that we get

$$x^6 - x^4 = 2a^2 + 2ax^2,$$

or

$$x^4 (2x^2 - 1) = (2a + x^2)^2,$$

which can be solved by the method indicated before.

Another equation is[3]

$$5x^4 = u^3.$$

In cases like this "the assumption should be always such," remarks Bhâskara II, "as will make it possible to remove (the cube of) the unknown." So assume $u = mx$; then

$$x = \tfrac{1}{5}m^3.$$

23. LINEAR FUNCTIONS MADE SQUARES OR CUBES

Square-pulveriser. The indeterminate equation of the type

$$bx + c = y^2$$

[1] *BBi*, pp. 103, 107. [2] *BBi*, p. 103; also *vide infra*, p. 280.

[3] *BBi*, p. 50; also *vide infra*, p. 278.

is called *varga-kuṭṭaka* or the "Square-pulveriser,"[1] inasmuch as, when expressed in the form

$$\frac{y^2 - c}{b} = x,$$

the problem reduces to finding a square (*varga*) which being diminished by c will be exactly divisible by b, which closely resembles the problem solved by the method of the pulveriser (*kuṭṭaka*).

For the solution in *integers* of an equation of this type, the method of the earlier writers appears to have been to assume suitable arbitrary values for y and then to solve the equation for x. Brahmagupta gives the following problems :

"The residue of the sun on Thursday is lessened and then multiplied by 5, or by 10 Making this (result) an exact square, within a year, a person becomes a mathematician."[2]

"The residue of any optional revolution lessened by 92 and then multiplied by 83 becomes together with unity a square. A person solving this within a year is a mathematician."[3]

That is to say, we are to solve the equations :

(1) $5x - 25 = y^2$,
(2) $10x - 100 = y^2$,
(3) $83x - 7635 = y^2$.

Pṛthûdakasvâmî solves them thus :

(1.1) Suppose $y = 10$; then $x = 125$. Or, put $y = 5$; then $x = 10$.

(2.1) Suppose $y = 10$; then $x = 20$.

(3.1) Assume $y = 1$; then $x = 92$.

[1] *BBi*, p. 122.
[2] *BrSpSi*, xviii 76.
[3] *BrSpSi*, xviii. 79.

He then remarks that by virtue of the multiplicity of suppositions there will be an infinitude of solutions in every case. But no method has been given either by Brahmagupta or by his commentator Pṛthûdakasvâmî to obtain the general solution.

The above method is reproduced by Bhāskara II.[1] He has also given the following rule:

"If a simple unknown be multiplied by the number which is the divisor of a square, etc., (on the other side) then, in order that its value may in such cases be integral, the square, etc., of another unknown should be put equal to (the other side). The rest (of the operations) will be as described before."[2]

His gloss on this rule runs as follows:

"In those cases, such as the Square-pulveriser, etc., where on taking the root of one side of the equation there remains on the other side a simple unknown multiplied by the number which was the divisor of the square, etc., the square, etc., of another unknown plus or minus an absolute term should be assumed for (the value of this other side) in order that its value may be integral. The rest (of the operations) will be as taught before."

Bhâskara has also quoted from an ancient author the following rule:

"(Find) a number whose square is exactly divisible by the divisor, as also its product by twice the square-root of the absolute term. An unknown multiplied by that number and superadded by the square-root of the absolute term should be assumed (for the unknown on the other side). If the absolute term does not yield a square-root, then after dividing it by the divisor, the

[1] *Vide infra*, p. 255f.

[2] *BBi*, p. 120.

remainder should be increased by so many times the divisor as will make a square. If this is not possible, then the problem is not soluble."[1]

Case i Let c be a square, equal to β^2, say. Then we have to solve

$$bx + \beta^2 = y^2.$$

The rule says, find p such that

$$p^2 = bq, \quad 2p\beta = br.$$

Then assume $y = pu + \beta$;

whence we get $x = qu^2 + ru$.

Bhâskara II prefers the assumption

$$y = bv + \beta,$$

so that we have $x = bv^2 + 2\beta v$.

Case ii. If c is not a square, suppose $c = bm + n$. Then find s such that

$$n + sb = r^2.$$

Now assume $y = bu \pm r$.

Substituting this value in the equation $bx + c = y^2$, we get

$$bx + c = (bu \pm r)^2$$
$$= b^2u^2 \pm 2bru + r^2,$$

or $$bx + c - r^2 = b^2u^2 \pm 2bru,$$

or $$bx + b(m - s) = b^2u^2 \pm 2bru.$$

$\therefore$ $$x = bu^2 \pm 2ru - (m - s).$$

Example from Bhâskara II :[2]

$$7x + 30 = y^2.$$

On dividing 30 by 7 the remainder is found to be 2; we also know that $2 + 7 \cdot 2 = 4^2$. Therefore, we

[1] *BBi*, p. 121. [2] *BBi*, pp. 120, 121.

assume in accordance with the above rule

$$y = 7u \pm 4;$$

whence we get $\qquad x = 7u^2 \pm 8u - 2,$

which is the general solution.

Cube-pulveriser. The indeterminate equation of the type

$$bx + c = y^3$$

is called the *ghana-kuṭṭaka* or the "Cube-pulveriser."[1] For its solution in integers Bhâskara II says:

"A method similar to the above may be applied also in the case of a cube thus: (find) a number whose cube is exactly divisible by the divisor, as also its product by thrice the cube-root of the absolute term. An unknown multiplied by that number and superadded by the cube-root of the absolute term should be assumed. If there be no cube-root of the absolute term, then after dividing it by the divisor, so many times the divisor should be added to the remainder as will make a cube. The cube-root of that will be the root of the absolute number. If there cannot be found a cube, even by so doing, that problem will be insoluble."[2]

Case i. Let $c = \beta^3$. Then we shall have to find p such that

$$p^3 = bq, \quad 3p\beta = br.$$

Now assume $\qquad y = pv + \beta.$

Substituting in the equation $bx + \beta^3 = y^3$ we get

$$\begin{aligned} bx + \beta^3 &= (pv + \beta)^3 \\ &= p^3v^3 + 3pv\beta(pv + \beta) + \beta^3, \end{aligned}$$

or $$bx = bqv^3 + brv(pv + \beta).$$

$\therefore$ $$x = qv^3 + rv(pv + \beta).$$

[1] *BBi*, p. 122. [2] *BBi*, p. 121.

Or, if we assume $y = bv + \beta$, we shall have

$$x = b^2v^3 + 3\beta v(bv + \beta).$$

Case ii. $c \neq$ a cube. Suppose $c = bm + n$; then find s such that

$$n + sb = r^3.$$

Now assume $y = bv + r$, whence we get

$$x = b^2v^3 + 3rv(bv + r) - (m - s),$$

as the general solution.

Example from Bhâskara II :[1]

$$5x + 6 = y^3.$$

Since $6 = 5 . 1 + 1$ and $1 + 43 \cdot 5 = 6^3$,

we assume $y = 5v + 6.$

Therefore $x = 25v^3 + 18v(5v + 6) + 42,$

is the general solution.

Equation $\mathbf{bx \pm c = ay^2}$. To solve an equation of the type

$$ay^2 = bx \pm c,$$

Bhâskara II says :

"Where the first side of the equation yields a root on being multiplied or divided[2] (by a number), there also the divisor will be as stated in the problem but the absolute term will be as modified by the operations."[3]

[1] *BBi*, p. 122.

[2] The printed text has *hitvâ kṣiptâ* (subtracting or adding). After collating with several copies Colebrooke accepted the reading *hatvâ kṣiptâ* (multiplying or adding). But we think that the correct reading should be *hatvâ hṛtvâ* (multiplying or dividing) For in his gloss Bhâskara II has employed the terms *guṇito vibhakto va* (multiplied or divided). Our emendation is further supported by the *rationale* of the rule.

[3] *BBi*, p. 121.

What is implied is this : Multiplying both sided of the given equation by a, we get

$$a^2y^2 = abx \pm ac,$$

Put $u = ay$, $v = ax$. Then the equation reduces to

$$u^2 = bv \pm ac,$$

which can be solved in the way described before.

We take the following illustrative example with its solution from Bhâskara II :[1]

$$5y^2 + 3 = 16x.$$

Multiplying by 5, and putting $u = 5y$, $v = 5x$, we get

$$u^2 = 16v - 15.$$

The solution of this is

$$u = 8w \pm 1,$$
$$v = 4w^2 \pm w + 1.$$

Therefore, we have

$$(1) \quad 5y = 8w + 1,$$

or $$(2) \quad 5y = 8w - 1.$$

Now, solving by the method of the pulveriser, we get the solution of (1) as

$$y = 8t + 5,$$
$$w = 5t + 3;$$

and that of (2) as

$$y = 8t + 3,$$
$$w = 5t + 2;$$

where t is any rational number.

Equation $bx \pm c = ay^n$ After describing the above methods Bhâskara II observes, *ityagre'pi yojyamiti śeṣaḥ* or "the same method can be applied further on

[1] *BBi*, p. 121.

(*i.e.*, to the cases of higher powers)"[1] Again at the end of the section he has added *evaṁ buddhimadbhiranyadapi yathâsambhavaṁ yojyam*, *i e*, "similar devices should be applied by the intelligent to further cases as far as practicable."[2] What is implied is as follows:

(1) To solve $\frac{x^n \pm c}{b} = y$.

Put $x = mz \pm k$. Then

$$\frac{x^n \pm c}{b} = \frac{1}{b}\left\{m^n z^n \pm nm^{n-1}z^{n-1}k + \frac{n(n-1)}{2}m^{n-2}z^{n-2}k^2 \pm \dots + nmz(\pm k)^{n-1} + (\pm k)^n \pm c\right\}$$

$$= \frac{1}{b}\left\{m^n z^n \pm nm^{n-1}z^{n-1}k + \dots + nmz(\pm k)^{n-1}\right\} + (\pm)^n\left(\frac{k^n \pm c}{b}\right)$$

Now, if

$$\frac{k^n \pm c}{b} = \text{a whole number},$$

$\frac{x^n \pm c}{b}$ will be an integral number when (1) $m = b$ or (2) b is a factor of m^n, $nm^{n-1}k$, etc. Or, in other words, knowing one integral solution of (1) an infinite number of others can be derived.

(2) To solve $\frac{ax^n \pm c}{b} = y$.

Multiplying by a^{n-1}, we get

$$\frac{a^n x^n \pm ca^{n-1}}{b} = ya^{n-1}.$$

[1] *BBi*, p. 121. [2] *BBi*, p. 122.

Putting $u = ax$, $v = ya^{n-1}$, we have

$$\frac{u^n \pm ca^{n-1}}{b} = v,$$

which is similar to case (1).

24. DOUBLE EQUATIONS OF THE FIRST DEGREE

The earliest instance of the solution of the simultaneous indeterminate quadratic equation of the type

$$\left.\begin{array}{l} x \pm a = u^2, \\ x \pm b = v^2, \end{array}\right\}$$

is found in the Bakhshâlî treatise. The portion of the manuscript containing the rule is mutilated. The example given in illustration can, however, be restored as follows :

"A certain number being added by five {becomes capable of yielding a square-root} ; the same number {being diminished by} seven becomes capable of yielding a square-root. What is that number is the question."[1]

That is to say, we have to solve

$$\sqrt{x+5} = u, \qquad \sqrt{x-7} = v.$$

The solution given is as follows :

"The sum of the additive and subtractive is |12|; its half |6|; minus two |4|; its half is |2|; squared |4|. 'Should be increased by the subtractive'; {the subtractive is} |7|; adding this we get |11|. This is the number (required)."

From this it is clear that the author gives the

[1] *BMs*, Folio 59, recto.

solution of the equations

$$x + a = u^2, \quad x - b = v^2;$$

as
$$x = \left\{\tfrac{1}{2}\left(\frac{a+b}{m} - m\right)\right\}^2 + b,$$

where m is any integer.[1]

Brahmagupta (628) gives the solution of the general case. He says:

"The difference of the two numbers by the addition or subtraction of which another number becomes a square, is divided by an optional number and then increased or decreased by it. The square of half the result diminished or increased by the greater or smaller (of the given numbers) is the number (required)."[2]

i.e.,
$$x = \left\{\tfrac{1}{2}\left(\frac{a-b}{m} \pm m\right)\right\}^2 \mp a,$$

or
$$x = \left\{\tfrac{1}{2}\left(\frac{a-b}{m} \mp m\right)\right\}^2 \mp b,$$

where m is an arbitrary integer.

The *rationale* is very simple. Since

$$u^2 = x \pm a, \quad v^2 = x \pm b,$$

we have
$$u^2 - v^2 = \pm a \mp b.$$

Therefore
$$u - v = m,$$

and
$$u + v = \frac{\pm a \mp b}{m},$$

where m is arbitrary. Hence

$$u = \tfrac{1}{2}\left(\frac{\pm a \mp b}{m} + m\right) = \pm \tfrac{1}{2}\left(\frac{a-b}{m} \pm m\right).$$

[1] In the above solution m is taken to be 2.

[2] *BrSpSi*, xviii. 74.

Since it is obviously immaterial whether u is taken as positive or negative, we have

$$u = \tfrac{1}{2}\left(\frac{a-b}{m} \pm m\right).$$

Similarly
$$v = \tfrac{1}{2}\left(\frac{a-b}{m} \mp m\right).$$

Therefore
$$x = \left\{\tfrac{1}{2}\left(\frac{a-b}{m} \pm m\right)\right\}^2 \mp a,$$

or
$$x = \left\{\tfrac{1}{2}\left(\frac{a-b}{m} \mp m\right)\right\}^2 \mp b,$$

where m is an arbitrary number.

Brahmagupta gives another rule for the particular case:

$$x + a = u^2,$$
$$x - b = v^2.$$

"The sum of the two numbers the addition and subtraction of which make another number (severally) a square, is divided by an optional number and then diminished by that optional number. The square of half the remainder increased by the subtractive number is the number (required)."[1]

i.e.,
$$x = \left\{\tfrac{1}{2}\left(\frac{a+b}{m} - m\right)\right\}^2 + b.$$

Nârâyaṇa (1357) says:

"The sum of the two numbers by which another number is (severally) increased and decreased so as to make it a square is divided by an optional number and then diminished by it and halved; the square of the result added with the subtrahend is the other number."[2] He further states:

[1] *BrSpSi*, xviii. 73. [2] *GK*, i. 52.

"The difference of the two numbers by which another number is increased twice so as to make it a square (every time), is increased by unity and then halved. The square of the result diminished by the greater number is the other number."[1]

i.e.,
$$x = \left(\frac{a - b + 1}{2}\right)^2 - a$$

is a solution of
$$x + a = u^2,\ x + b = v^2,\ a > b.$$

"The difference of the two numbers by which another number is diminished twice so as to make it a square (every time), is decreased by unity and then halved. The result multiplied by itself and added with the greater number gives the other."[2]

i.e.,
$$x = \left(\frac{a - b - 1}{2}\right)^2 + a$$

is a solution of
$$x - a = u^2,\ x - b = v^2,\ a > b.$$

The general case
$$\left.\begin{aligned} ax + c = u^2, \\ bx + d = v^2, \end{aligned}\right\} \qquad (1)$$

has been treated by Bhâskara II. He first lays down the rule :

"In those cases where remains the (simple) unknown with an absolute number, there find its value by forming an equation with the square, etc., of another unknown plus an absolute number. Then proceed to the solution of the next equation comprising the simple unknown with an absolute number by substituting in it the root obtained before."[3]

[1] *GK*, i. 53.
[2] *GK*, i. 54.
[3] *BBi*, pp. 117-8.

He then proceeds to explain it further :

"In those cases where on taking the square-root of the first side, there remains on the other side the (simple) unknown with or without an absolute number, find there the value of that unknown by forming an equation with the square of another unknown plus an absolute number. Having obtained the value of the unknown in this way and substituting that value (in the next equation) further operations should be proceeded with. If, however, on equating the root of the first with another unknown plus an absolute number, no further operations remain to be done, then the equation has to be made with the square, etc., of a known number."

(*i*) Set $u = mw + \alpha$; then substituting in the first equation, we get

$$x = \frac{1}{a}(m^2w^2 + 2mw\alpha + \alpha^2 - c).$$

Substituting this value of x in the next equation, we have

$$\frac{b}{a}(m^2w^2 + 2mw\alpha + \alpha^2 - c) + d = v^2, \qquad (1.1)$$

which can be solved by the method of the Square-nature.

(*ii*) In the course of working out an example[1] Bhâskara II is found to have followed also a different procedure, which was subsequently adopted by Lagrange.[2]

Eliminate x between the two equations. Then

$$bu^2 + (ad - bc) = av^2,$$

or $$abu^2 + k = w^2, \qquad (1.2)$$

where $w = av$, $k = a^2d - abc$.

[1] *Vide infra*, p. 265.
[2] *Addition to Euler's Algebra*, p. 547.

If $u = r$, $w = s$ be a solution of this transformed equation, another solution of it will be

$$u = rq \pm ps,$$
$$w = qs \pm abrp;$$

where $abp^2 + 1 = q^2$. Therefore, the general solution of (1) is

$$x = \frac{1}{a}(rq \pm ps)^2 - \frac{c}{a},$$
$$u = rq \pm ps,$$
$$v = \frac{1}{a}(qs \pm abrp);$$

where $abp^2 + 1 = q^2$ and $abr^2 + a^2d - abc = s^2$.

Now, a rational solution of the equation $abp^2 + 1 = q^2$ is

$$p = \frac{2t}{t^2 - ab}, \quad q = \frac{t^2 + ab}{t^2 - ab},$$

where t is any rational number. Therefore, the above general solution becomes

$$\left.\begin{aligned} x &= \frac{1}{a(t^2 - ab)^2}\left\{ r(t^2 + ab) \pm 2st \right\}^2 - \frac{c}{a}, \\ u &= \frac{1}{(t^2 - ab)}\left\{ r(t^2 + ab) \pm 2st \right\}, \\ v &= \frac{1}{a(t^2 - ab)}\left\{ s(t^2 + ab) \pm 2abrt \right\}, \end{aligned}\right\} \quad (1 \cdot 3)$$

where $abr^2 + a^2d - abc = s^2$.

(*iii*) Suppose c and d to be squares, so that $c = \alpha^2$, $d = \beta^2$. Then we shall have to solve

$$ax + \alpha^2 = u^2,$$
$$bx + \beta^2 = v^2.$$

The auxiliary equation $abr^2 + a^2d - abc = s^2$ in this case becomes

$$abr^2 + (a^2\beta^2 - ab\alpha^2) = s^2.$$

The same equation is obtained by proceeding as in case (*i*) with the assumption $v = by + \beta$.

An obvious solution of it is $r = \alpha$, $s = a\beta$. Hence in this case the general solution (1.3) becomes

$$x = \frac{1}{a(t^2 - ab)^2}\{\alpha(t^2 + ab) \pm 2a\beta t\}^2 - \frac{\alpha^2}{c},$$

$$u = \frac{1}{(t^2 - ab)}\{\alpha(t^2 + ab) \pm 2a\beta t\},$$

$$v = \frac{1}{(t^2 - ab)}\{\beta(t^2 + ab) \pm 2b\alpha t\},$$

where t is any rational number.

Putting $\alpha = \beta = 1$, $t = a$, and taking the positive sign only, we get a particular solution of the equations

$$\left.\begin{aligned} ax + 1 &= u^2 \\ bx + 1 &= v^2 \end{aligned}\right\}$$

as

$$x = \frac{8(a + b)}{(a - b)^2}, \quad u = \frac{3a + b}{a - b}, \quad v = \frac{a + 3b}{a - b}.$$

This solution has been stated by Brahmagupta (628):

"The sum of the multipliers multiplied by 8 and divided by the square of the difference of the multipliers is the (unknown) number. Thrice the two multipliers increased by the alternate multiplier and divided by their difference will be the two roots."[1]

It has been described partly by Nârâyaṇa (1357) thus :

[1] *BrSpSi*, xviii. 71.

"The two numbers by which another number is multiplied at two places so as to make it (at every place), together with unity, a square, their sum multiplied by 8 and divided by the square of their difference is the other number."[1]

We take an illustrative example with its solution from Bhâskara II :

"If thou be expert in the method of the elimination of the middle term, tell the number which being severally multiplied by 3 and 5, and then added with unity, becomes a square."[2]

That is to say, we have to solve

$$\left.\begin{aligned} 3x + 1 &= u^2, \\ 5x + 1 &= v^2. \end{aligned}\right\}$$

Bhâskara II solves these equations substantially as follows :

(1) Set $u = 3y + 1$; then from the first equation,

$$x = 3y^2 + 2y.$$

Substituting this value in the other equation, we get

$$15y^2 + 10y + 1 = v^2,$$

or $$(15y + 5)^2 = 15v^2 + 10.$$

By the method of the Square-nature we have the solutions of this equation as

$$\left.\begin{aligned} v &= 9 \\ 15y + 5 &= 35 \end{aligned}\right\}, \quad \left.\begin{aligned} v &= 71 \\ 15y + 5 &= 275 \end{aligned}\right\}\ldots,$$

Therefore $y = 2, 18, \ldots$

Hence $x = 16, 1008, \cdots$

(2) Or assume the unknown number to be

$$x = \tfrac{1}{3}(u^2 - 1),$$

[1] *GK*, i. 51. [2] *BBi*, p. 118.

so that the first condition of the problem (*i.e.*, the first equation) is identically statisfied. Then by the second condition

$$\tfrac{5}{3}(u^2 - 1) + 1 = v^2,$$

or
$$(5u)^2 = 15v^2 + 10.$$

Now, by the method of the Square-nature, we get the values of (u, v) as (7, 9), (55, 71), etc. Therefore, the values of x are, as before, 16, 1008, etc.

The following example is by Nârâyaṇa:

"O expert in the art of the Square-nature, tell me the number which being severally multiplied by 4 and 7 and decreased by 3, becomes capable of yielding a square-root."[1]

That is, solve:

$$\left.\begin{aligned} 4x - 3 &= u^2, \\ 7x - 3 &= v^2. \end{aligned}\right\}$$

Nârâyaṇa says: "By the method indicated before the number is 1, 21, or 1057."

25. DOUBLE EQUATIONS OF THE SECOND DEGREE

First Type. The double equations of the second degree considered by the Hindus are of two general types. The first of them is

$$\left.\begin{aligned} ax^2 + by^2 + c &= u^2, \\ a'x^2 + b'y^2 + c' &= v^2. \end{aligned}\right\}$$

Of these the more thoroughly treated particular cases are as follows:

Case i.
$$\begin{cases} x^2 + y^2 + 1 = u^2, \\ x^2 - y^2 + 1 = v^2, \end{cases}$$

[1] *GK*, p. 40.

Case ii. $\begin{cases} x^2 + y^2 - 1 = u^2, \\ x^2 - y^2 - 1 = v^2. \end{cases}$

It should be noted that though the earliest treatment of these equations is now found in the algebra of Bhâskara II (1150), they have been admitted by him as being due to previous authors (*âdyodâharaṇam*).

For the solution of (*i*) Bhâskara II assumes[1]

$$x^2 = 5z^2 - 1, \quad y^2 = 4z^2,$$

so that both the equations are satisfied. Now, by the method of the Square-nature, the solutions of the equation $5z^2 - 1 = x^2$ are (1, 2), (17, 38),...Therefore, the solutions of (*i*) are

$$\left.\begin{matrix} x = 2 \\ y = 2 \end{matrix}\right\}, \quad \left.\begin{matrix} x = 38 \\ y = 34 \end{matrix}\right\}, \quad \ldots$$

Similarly, for the solution of (*ii*), he assumes

$$x^2 = 5z^2 + 1, \quad y^2 = 4z^2,$$

which satisfy the equations. By the method of the Square-nature the values of (z, x) in the equation $5z^2 + 1 = x^2$ are (4, 9), (72, 161), etc. Hence the solutions of (*ii*) are

$$\left.\begin{matrix} x = 9 \\ y = 8 \end{matrix}\right\} \quad \left.\begin{matrix} x = 161 \\ y = 144 \end{matrix}\right\}, \quad \ldots$$

Bhâskara II further says that for the solution of equations of the forms (*i*) and (*ii*) a more general assumption will be

$$x^2 = pz^2 \mp 1, \quad y^2 = m^2z^2;$$

where p, m are such that

$$p \pm m^2 = \text{a square},$$

[1] *BBi*, p. 108.

the upper sign being taken for Case *i* and the lower sign for Case *ii*. Both the equations are then identically satisfied. Suppose

$$p + m^2 = s^2,$$
$$p - m^2 = t^2.$$

Whence
$$s = \tfrac{1}{2}\left(\frac{2m^2}{n} + n\right),$$

$$t = \tfrac{1}{2}\left(\frac{2m^2}{n} - n\right),$$

where n is any rational number. Therefore

$$p = \tfrac{1}{4}\left(\frac{4m^4}{n^2} + n^2\right).$$

Here he observes that m^2 should be so chosen that p will be an integer.

Hence
$$\left.\begin{aligned} x^2 &= \tfrac{1}{4}\left(\frac{4m^4}{n^2} + n^2\right)z^2 \mp 1, \\ y^2 &= m^2 z^2; \end{aligned}\right\} \qquad (1)$$

the upper sign being taken for Case *i* and the lower sign for Case *ii*.

Whence
$$u = \tfrac{1}{2}\left(\frac{2m^2}{n} + n\right)z,$$

$$v = \tfrac{1}{2}\left(\frac{2m^2}{n} - n\right)z.$$

Or, we may proceed in a different way, says Bhâskara II:

Since

$$(p^2 + q^2) \pm 2pq$$

is always a square, we may assume

$$x^2 = (p^2 + q^2)w^2 \mp 1,$$
$$y^2 = 2pqw^2.$$

For a rational value of y, $2pq$ must be a square. So we take

$$p = 2m^2, \; q = n^2.$$

Hence we have the assumption

$$\left.\begin{aligned} x^2 &= (4m^4 + n^4)w^2 \mp 1, \\ y^2 &= 4m^2n^2w^2; \end{aligned}\right\} \qquad (2)$$

the upper sign being taken for Case *i* and the lower sign for Case *ii*.

Whence

$$u = (2m^2 + n^2)w,$$
$$v = (2m^2 - n^2)w.$$

It will be noticed that the equations (1) follow from (2) on putting $w = z/2n$. So we shall take the latter as our fundamental assumption for the solution of the equations (*i*) and (*ii*). Then, from the solutions of the subsidiary equations

$$(4m^4 + n^4)w^2 \mp 1 = x^2,$$

by the method of the Square-nature, observes Bhâskara II, an infinite number of integral solutions of the original equations can be derived.[1]

Now, one rational solution of

$$(4m^4 + n^4)w^2 + 1 = x^2$$

is

$$w = \frac{2r}{(4m^4 + n^4) - r^2},$$

$$x = \frac{(4m^4 + n^4) + r^2}{(4m^4 + n^4) - r^2}.$$

Therefore, we have the general solution of

$$\left.\begin{aligned} x^2 + y^2 - 1 &= v^2 \\ x^2 - y^2 - 1 &= u^2 \end{aligned}\right\}$$

[1] Cf. *BBi*, p. 110.

as

$$\left.\begin{aligned} x &= \frac{(4m^4 + n^4) + r^2}{(4m^4 + n^4) - r^2}, \quad u = \frac{2r(2m^2 + n^2)}{(4m^4 + n^4) - r^2}, \\ y &= \frac{4mnr}{(4m^4 + n^4) - r^2}, \quad v = \frac{2r(2m^2 - n^2)}{(4m^4 + n^4) - r^2}; \end{aligned}\right\} \quad (A)$$

where m, n, r are rational numbers.

For $r = s/t$, we get Genocchi's solution.[1]

In particular, put $m = 2t$, $n = 1$, $r = 8t^2 - 1$ in (A). Then, we get the solution

$$\left.\begin{aligned} x &= \tfrac{1}{2}\left(\frac{8t^2 - 1}{2t}\right)^2 + 1, \quad u = \frac{64t^4 - 1}{8t^2}, \\ y &= \frac{8t^2 - 1}{2t}, \quad v = \tfrac{1}{2}\left(\frac{8t^2 - 1}{2t}\right)^2 \end{aligned}\right\} \quad (a)$$

Putting $m = t$, $n = 1$, $r = 2t^2 + 2t + 1$ in (A), we have[2]

$$\left.\begin{aligned} x &= t + \frac{1}{2t}, \quad u = t + \frac{1}{2t}, \\ y &= 1, \quad v = t - \frac{1}{2t}. \end{aligned}\right\} \quad (b)$$

Again, if we put $m = t$, $n = 1$, $r = 2t^2$ in (A), we get

$$\left.\begin{aligned} x &= 8t^4 + 1, \quad u = 4t^2(2t^2 + 1), \\ y &= 8t^3, \quad v = 4t^2(2t^2 - 1). \end{aligned}\right\} \quad (c)$$

These three solutions have been stated by Bhâskara II in his treatise on arithmetic. He says,

[1] *Nouv. Ann. Math.*, X, 1851, pp. 80-85; also Dickson, *Numbers*, II, pp. 479. For a summary of important Hindu results in algebra see the article of A. N. Singh in the *Archeon*, 1936.

[2] Here, and also in (c), we have overlooked the negative sign of x, y, u and v.

"The square of an optional number is multiplied by 8, decreased by unity, halved and then divided by that optional number. The quotient is one number. Half its square plus unity is the other number. Again, unity divided by twice an optional number added with that optional number is the first number and unity is the second number. The sum and difference of the squares of these two numbers minus unity will be (severally) squares."[1]

"The biquadrate and the cube of an optional number is multiplied by 8, and the former product is again increased by unity. The results will be the two numbers (required)."[2]

Nârâyaṇa writes :

"The cube of any optional number is the first number; half the square of its square plus unity is the second. The sum and difference of the squares of these two numbers minus unity become squares."[3]

That is, if m be an optional number, one solution of (*ii*), according to Nârâyaṇa, is

$$x = \frac{m^4}{2} + 1, \qquad u = (m^2 + 2)\frac{m^2}{2},$$

$$y = m^3, \qquad v = (m^2 - 2)\frac{m^2}{2}.$$

It will be noticed that this solution follows easily from the solution (*c*) of Bhâskara II, on putting $t = m/2$. This special solution was found later on by E. Clere (1850).[4]

[1] *L*, p. 13. [2] *L*, p. 14.
[3] *GK*, i. 46.
[4] *Nouv. Ann. Math.*, IX, 1850, pp. 116-8; also Dickson, *Numbers*, II, p. 479; Singh, l. c.

Now, let us take into consideration the equation

$$(4m^4 + n^4)w^2 - 1 = x^2.$$

Its solutions are known to be

$$\left.\begin{aligned} w &= \frac{1}{n^2} \\ x &= \frac{2m^2}{n^2} \end{aligned}\right\}, \qquad \left.\begin{aligned} w &= \frac{1}{2m^2} \\ x &= \frac{n^2}{2m^2} \end{aligned}\right\}$$

From these, by the Principle of Composition, we get respectively two other solutions

$$\left.\begin{aligned} w &= \frac{16m^4 + n^4}{n^6} \\ x &= \frac{32m^6 + 6m^2n^4}{n^6} \end{aligned}\right\}, \qquad \left.\begin{aligned} w &= \frac{m^4 + n^4}{2m^6} \\ x &= \frac{n^6 + 3n^2m^4}{2m^6} \end{aligned}\right\}$$

Therefore, the general solutions of

$$\left.\begin{aligned} x^2 + y^2 + 1 &= u^2, \\ x - y^2 + 1 &= v^2; \end{aligned}\right\}$$

are

$$\left.\begin{aligned} x &= \frac{2m^2}{n^2}, & u &= \frac{2m^2 + n^2}{n^2}, \\ y &= \frac{2m}{n}, & v &= \frac{2m^2 - n^2}{n^2}; \end{aligned}\right\} \qquad (a')$$

$$\left.\begin{aligned} x &= \frac{1}{n^6}(32m^6 + 6m^2n^4), \\ y &= \frac{2m}{n^5}(16m^4 + n^4), \\ u &= \frac{1}{n^6}(16m^4 + n^4)(2m^2 + n^2), \\ v &= \frac{1}{n^6}(16m^4 + n^4)(2m^2 - n^2); \end{aligned}\right\} \qquad (a'')$$

$$\left.\begin{aligned} x &= \frac{n^2}{2m^2}, & u &= \frac{2m^2+n^2}{2m^2}, \\ y &= \frac{n}{m}, & v &= \frac{2m^2-n^2}{2m^2}; \end{aligned}\right\} \qquad (b')$$

and

$$\left.\begin{aligned} x &= \frac{1}{2m^6}(n^6+3n^2m^4), \\ y &= \frac{n}{m^5}(m^4+n^4), \\ u &= \frac{1}{2m^6}(m^4+n^4)(2m^2+n^2), \\ v &= \frac{1}{2m^6}(m^4+n^4)(2m^2-n^2). \end{aligned}\right\} \qquad (b'')$$

Putting $n = 1$ in (a') and (a''), we have the integral solutions

$$\left.\begin{aligned} x &= 2m^2, & u &= 2m^2+1, \\ y &= 2m, & v &= 2m^2-1; \end{aligned}\right\} \qquad (a'.1)[1]$$

and

$$\left.\begin{aligned} x &= 2m^4(16m^2+3), \\ y &= 2m(16m^4+1), \\ u &= (16m^4+1)(2m^2+1), \\ v &= (16m^4+1)(2m^2-1). \end{aligned}\right\} \qquad (a''.1)$$

Similarly, if we put $m = 1$ in (b') and (b''), we get

$$\left.\begin{aligned} x &= \tfrac{1}{2}n^2, & u &= \tfrac{1}{2}(n^2+2), \\ y &= n, & v &= \tfrac{1}{2}(n^2-2); \end{aligned}\right\} \qquad (b'.1)$$

and

$$\left.\begin{aligned} x &= \tfrac{1}{2}n^2(n^4+3), & u &= \tfrac{1}{2}(n^4+1)(n^2+2), \\ y &= n(n^4+1), & v &= \tfrac{1}{2}(n^4+1)(n^2-2). \end{aligned}\right\} \qquad (b''.1)$$

[1] This solution was given by Drummond (*Amer. Math. Mon.*, IX, 1902, p. 232).

The solution (b'. 1) is stated by Nârâyaṇa thus:

"Any optional number is the first and half its square is the second. The sum and difference of the squares of these two numbers with unity become capable of yielding a square-root."[1]

Case iii. Form

$$\left.\begin{aligned} ax^2 + by^2 &= u^2, \\ a'x^2 + b'y^2 + c' &= v^2. \end{aligned}\right\}$$

For the solution of double equations of this form Bhâskara II adopts the following method:

The solution of the first equation is $x = my$, $u = ny$; where

$$am^2 + b = n^2.$$

Substituting in the second equation, we get

$$(a'm^2 + b')y^2 + c' = v^2,$$

which can be solved by the method of the Square-nature.

Example from Bhâskara II:[2]

$$\left.\begin{aligned} 7x^2 + 8y^2 &= u^2 \\ 7x^2 - 8y^2 + 1 &= v^2 \end{aligned}\right\}.$$

He solves it substantially as follows:

In the first equation suppose $x = 2y$; then $u = 6y$. Putting $x = 2y$, the second equation becomes

$$20y^2 + 1 = v^2.$$

By the method of the Square-nature the values of y satisfying this equation are 2, 36, etc. Hence the solutions of the given double equation are

$$x = 4, 72, \ldots$$

$$y = 2, 36, \ldots$$

[1] *GK*, i. 45.

[2] *BBi*, p. 119.

Case iv. Form

$$\left.\begin{aligned} a(x^2 \pm y^2) + c &= u^2, \\ a'(x^2 \pm y^2) + c' &= v^2. \end{aligned}\right\}$$

Putting $x^2 \pm y^2 = z$ Bhâskara II reduces the above equations to

$$\left.\begin{aligned} az + c &= u^2, \\ a'z + c' &= v^2; \end{aligned}\right\}$$

the method for the solution of which has been given before.

Example with solution from Bhâskara II :[1]

$$\left.\begin{aligned} 2(x^2 - y^2) + 3 &= u^2 \\ 3(x^2 - y^2) + 3 &= v^2 \end{aligned}\right\}.$$

Set $x^2 - y^2 = z$; then

$$2z + 3 = u^2,$$

$$3z + 3 = v^2.$$

Eliminating z we get

$$3u^2 = 2v^2 + 3,$$

or $$(3u)^2 = 6v^2 + 9.$$

Whence $$v = 6, 60, \ldots$$

$$3u = 15, 147, \ldots$$

Therefore $$u = 5, 49, \ldots$$

Hence $$x^2 - y^2 = z = 11, 1199, \ldots$$

Therefore, the required solutions are

$$\left.\begin{aligned} x &= \tfrac{1}{2}\left(\frac{11}{m} + m\right) \\ y &= \tfrac{1}{2}\left(\frac{11}{m} - m\right) \end{aligned}\right\}, \quad \left.\begin{aligned} x &= \tfrac{1}{2}\left(\frac{1199}{m} + m\right) \\ y &= \tfrac{1}{2}\left(\frac{1199}{m} - m\right) \end{aligned}\right\}, \ \ldots$$

where m is an arbitrary rational number.

[1] *BBi*, p. 119.

For $m = 1$, the values of (x, y) will be $(6, 5)$, $(600, 599), \ldots$

For $m = 11$, we get the solution $(60, 49), \cdots$

Case v. For the solution of the double equation of the general form

$$\left.\begin{array}{r} ax^2 + by^2 + c = u^2 \\ a'x^2 + b'y^2 + c' = v^2 \end{array}\right\}$$

Bhâskara II's hint[1] is : Find the values of x, u in the first equation in terms of y, and then substitute that value of x in the second equation so that it will be reduced to a Square-nature. He has, however, not given any illustrative example of this kind.

Second Type. Another type of double equation of the second degree which has been treated is

$$\left.\begin{array}{r} a^2x^2 + bxy + cy^2 = u^2, \\ a'x^2 + b'xy + c'y^2 + d' = v^2. \end{array}\right\}$$

The solution of the first equation has been given before to be

$$x = \frac{1}{2a}\left\{\frac{y^2}{\lambda}\left(c - \frac{b^2}{4a^2}\right) - \lambda\right\} - \frac{by}{2a^2},$$

$$u = \tfrac{1}{2}\left\{\frac{y^2}{\lambda}\left(c - \frac{b^2}{4a^2}\right) + \lambda\right\},$$

where λ is an arbitrary rational number. Putting $\lambda = y$, we have

$$x = \frac{y}{2a}\left(c - \frac{b^2}{4a^2} - 1\right) - \frac{by}{2a^2} = \alpha y,$$

$$u = \frac{y}{2}\left(c - \frac{b^2}{4a^2} + 1\right);$$

where $$\alpha = \frac{1}{2a}\left(c - \frac{b^2}{4a^2} - 1\right) - \frac{b}{2a^2}.$$

[1] *Vide supra*, pp. 190f.

Substituting in the second equation, we get

$$(a'\alpha^2 + b'\alpha + c')y^2 + d' = v^2,$$

which can be solved by the method of the Square-nature. This method is equally applicable if the unknown part in the second equation is of a different kind but still of the second degree.

Bhâskara II gives the following illustrative example together with its solution :[1]

$$\left.\begin{aligned} x^2 + xy + y^2 &= u^2 \\ (x+y)u + 1 &= v^2 \end{aligned}\right\}.$$

Multiplying the first equation by 36, we get

$$(6x + 3y)^2 + 27y^2 = 36u^2.$$

Whence
$$6x + 3y = \tfrac{1}{2}\left(\frac{27y^2}{\lambda} - \lambda\right),$$

$$6u = \tfrac{1}{2}\left(\frac{27y^2}{\lambda} + \lambda\right),$$

where λ is an arbitrary rational number. Taking $\lambda = y$, we have

$$6x + 3y = 13y,$$

or
$$x = \tfrac{5}{3}y,$$

and
$$u = \tfrac{7}{3}y.$$

Substituting in the second equation, we get

$$56y^2 + 9 = 9v^2.$$

By the method of the Square-nature the values of y are 6, 180, ...

Hence the required values of (x, y) are (10, 6), (300, 180), ...

[1] *BBi*, pp. 107f.

26. DOUBLE EQUATIONS OF HIGHER DEGREES

There are a few problems which involve double equations of degrees higher than the second. The following examples are taken from Bhâskara II:

Example 1. "The sum of the cubes (of two numbers) is a square and the sum of their squares is a cube. If you know them, then I shall admit that you are a great algebraist."[1]

We have to solve the equations

$$\left.\begin{aligned} x^2 + y^2 &= u^3, \\ x^3 + y^3 &= v^2. \end{aligned}\right\}$$

The solution of this problem by Bhâskara II is as follows:

"Here suppose the two numbers to be z^2, $2z^2$. The sum of their cubes is $9z^6$. This is by itself a square and its square-root is $3z^3$.

"Now the sum of the squares of those two numbers is $5z^4$. This must be a cube. Assuming it to be equal to the cube of an optional multiple of $5z$ and removing the factor z^3 from both sides (we get $z = 25p^3$, where p is an optional number); so, as stated before, the numbers are (putting $p = 1$) 625, 1250. The assumption should be always such as will make it possible to remove (the cube of) the unknown."[2]

In general, assume $x = mz^2$, $y = nz^2$; substituting in the second equation, we have

$$x^3 + y^3 = (m^3 + n^3)z^6 = v^2.$$

If $\quad m^3 + n^3 = \text{a square} = p^2$, say,

then $\quad v = pz^3.$

[1] *BBi*, p. 56. [2] *BBi*, pp. 56f.

Now, from the first equation, we have

$$(m^2 + n^2)z^4 = u^3.$$

Assume $u = rz$; then

$$z = \frac{r^3}{m^2 + n^2}.$$

Hence we get

$$x = \frac{mr^6}{(m^2 + n^2)^2}, \quad y = \frac{nr^6}{(m^2 + n^2)^2};$$

where r is any integer and m, n are such that

$$m^3 + n^3 = \text{a square.}$$

One obvious solution[1] of this equation is $m = 1$, $n = 2$. Hence we get the solution

$$x = \frac{r^6}{25}, \quad y = \frac{2r^6}{25}.$$

This particular solution has been given by Nârâyaṇa, who says:

"The square of the cube of an optional number is the one and twice it is the other. These divided by 25 will be the two numbers, the sum of whose squares will

[1] Now $m^3 + n^3$ can be made a square by putting

$$m = (p^3 + q^3)p, \quad n = (p^3 + q^3)q,$$

so that

$$m^3 + n^3 = (p^3 + q^3)^4.$$

Hence the solution of our equation will be

$$x = \frac{pr^6}{(p^2 + q^2)^2 (p^3 + q^3)^3},$$

$$y = \frac{qr^6}{(p^2 + q^2)^2 (p^3 + q^3)^3}.$$

Putting $r = (p^2 + q^2)(p^3 + q^3)s$, we have the solution in positive integers as

$$x = p(p^2 + q^2)^4 (p^3 + q^3)^3 s^6,$$

$$y = q(p^2 + q^2)^4 (p^3 + q^3)^3 s^6;$$

where p, q, s are any integral numbers.

be a cube and the sum of whose cubes will be a square."[1] He then adds by way of illustration :

"With the optional number 1, we get the two numbers (1/25, 2/25); with 2, (64/25, 128/25); with 5, (625, 1250); with 1/2, (1/1600, 1/800); with 1/3, (1/18225, 2/18225). Thus by virtue of (the multiplicity of) the optional number many solutions can be found."

Example 2. "O most learned algebraist, tell me those various pairs of whole numbers whose difference is a square and the sum of whose squares is a cube."[2]

That is to say, solve in positive integers

$$\left.\begin{array}{l} y - x = u^2, \\ y^2 + x^2 = v^3. \end{array}\right\}$$

Bhâskara II's process of solving this problem is as follows :

"Let the two numbers be x, y. Putting their difference, $y - x$, equal to u^2, we get the value of x as $y - u^2$. Having thus found the value of x, the two numbers become $y - u^2, y$.

"The sum of their squares $= 2y^2 - 2yu^2 + u^4$. This is a cube. Making it equal to u^6 and transposing we get

$$u^6 - u^4 = 2y^2 - 2yu^2.$$

Multiplying both sides by 2 and superadding u^4, we get the square-root of the second side $= 2y - u^2$, and the first side $= 2u^6 - u^4$. Dividing out by u^4 (and putting w for $2y/u^2 - 1$), we get

$$2u^2 - 1 = w^2.$$

By the method of the Square-nature the roots of this equation are

$$u = 5, 29, \quad \ldots$$
$$w = 7, 41, \quad \ldots$$

[1] *GK*, i. 50.

[2] *BBi*, p. 103.

"Then by the rule, 'Or, if a biquadratic factor has been removed, the greater root should be multiplied by the square of the lesser root,'[1] we get

$$2y - 25 = 175,$$

or

$$2y - 841 = 34481.$$

Therefore

$$y = 100, 17661, \ldots$$

"Finding the respective values of the numbers, they are (75, 100), (16820, 17661), etc."

Example 3. "Bring out quickly those two numbers of which the sum of the cube (of one) and the square (of the other) becomes a square and whose sum also is a square."[2]

That is to say, solve

$$\begin{cases} x^3 + y^2 = u^2, & (1) \\ x + y = v^2. & (2) \end{cases}$$

This problem has been solved by Bhâskara II in two ways, which are substantially as follows:

First method. From (1) we get

$$u = \tfrac{1}{2}\left(\frac{x^3}{\lambda} + \lambda\right), \quad y = \tfrac{1}{2}\left(\frac{x^3}{\lambda} - \lambda\right),$$

where λ is an arbitrary number. Putting $\lambda = x$, we get

$$u = \tfrac{1}{2}(x^2 + x), \quad y = \tfrac{1}{2}(x^2 - x).$$

Substituting this value of y in (2), we get

$$x^2 + x = 2v^2,$$

or

$$(2x + 1)^2 = 8v^2 + 1.$$

[1] The reference is to the rule on p. 249.

[2] *BBi*, p. 107.

By the method of the Square-nature we have

$$\left.\begin{array}{r} v = 6 \\ 2x + 1 = 17 \end{array}\right\}, \quad \left.\begin{array}{r} v = 35 \\ 2x + 1 = 99 \end{array}\right\}, \ldots$$

Whence the values of (x, y) are $(8, 28)$, $(49, 1176)$, ...

Second Method. Assume $x = 2w^2$, $y = 7w^2$. Then

$$x + y = 9w^2 = (3w)^2.$$

So the equation (2) is satisfied. Now, substituting those values in (1) we get

$$8w^6 + 49w^4 = u^2,$$

or
$$w^4(8w^2 + 49) = u^2.$$

If
$$8w^2 + 49 = z^2,$$

then
$$u = zw^2.$$

Now the values of w making $8w^2 + 49$ a square are 2, 3, 7... . Hence the required numbers (x, y) are $(8, 28)$, $(18, 63)$, $(98, 343)$, ...

Example 4. "What is that number which multiplied by three and added with unity becomes a cube; the cube-root squared and multiplied by three becomes, together with unity, a square."[1]

That is to say, solve

$$\begin{cases} 3x + 1 = u^3, & (1) \\ 3u^2 + 1 = v^2. & (2) \end{cases}$$

It has been solved by Bhâskara II thus:

From (2), by the method of the Square-nature, we get the values of (u, v) as $(1, 2)$, $(4, 7)$, $(15, 26)$, ... Whence the values of x are 21, 3374/3, ...

[1] *BBi*, p. 119. This problem is admittedly taken by Bhâskara II from an earlier writer.

Alternatively[1] we assume $u = 3y + 1$; then from the equation (1) we get

$$x = 3y(3y^2 + 3y + 1).$$

Also from (2) we have

$$27y^2 + 18y + 4 = v^2$$
$$= (my - 2)^2, \text{ say.}$$

Hence $$y = \frac{18 + 4m}{m^2 - 27}.$$

Therefore, the required value of x is

$$x = 9\left(\frac{18 + 4m}{m^2 - 27}\right)^3 + 9\left(\frac{18 + 4m}{m^2 - 27}\right)^2 + 3\left(\frac{18 + 4m}{m^2 - 27}\right),$$

where m is a rational number greater than 5.

The first of the previous solutions is given by $m = 9$.

Double Equations in Several Unknowns. To solve a double equation involving several unknowns, Bhâskara II gives the following hints :

"When there are square and other powers of three or more unknowns, leaving out any two unknowns at pleasure, the values of others should be arbitrarily assumed and the roots investigated."[2]

For the case of a single equation, he says :

"But when there is only one equation, the roots should be determined as before after assuming optional values for all the unknowns except one."

27. MULTIPLE EQUATIONS

There are some very elegant problems in which three or more functions, linear or quadratic, of the unknowns have to be made squares or cubes. The

[1] See *BBi*, p. 121. [2] *BBi*, p. 106.

following example occurs in the *Laghu-Bhâskarîya* of Bhâskara I[1] (522):

Example 1. To find two numbers x and y such that the expressions $x + y$, $x - y$, $xy + 1$ are each a perfect square.

Brahmagupta gives the following solution:

"A square is increased and diminished by another. The sum of the results is divided by the square of half their difference. Those results multiplied (severally) by this quotient give the numbers whose sum and difference are squares as also their product together with unity."[2]

Thus the solution is:

$$x = P(m^2 + n^2),$$
$$y = P(m^2 - n^2),$$

where $P = \dfrac{(m^2 + n^2) + (m^2 - n^2)}{[\frac{1}{2}\{(m^2 + n^2) - (m^2 - n^2)\}]^2}$, m, n being any rational numbers.

Nârâyaṇa (1357) says:

"The square of the square of an optional number is set down at two places. It is decreased by the square (at one place) and increased (at another), and then doubled. The sum and difference of the results are squares and so also their product together with unity."[3]

That is,
$$x = 2(p^4 + p^2),$$
$$y = 2(p^4 - p^2),$$

where p is any rational number.

[1] *LBh*, viii. 17. [2] *BrSpSi*, xviii. 72.
[3] *GK*, i. 47.

The *rationale* of this solution is as follows:

Suppose

$$x = 2z^2(m^2+n^2), \quad y = 2z^2(m^2-n^2),$$

so that $x \pm y$ are already squares. We have, therefore, only to make

$$xy + 1 = 4z^4(m^2+n^2)(m^2-n^2) + 1 = \text{a square.}$$

Now

$$4z^4(m^4-n^4) + 1 = (2z^2m^2-1)^2 + 4z^2(m^2-z^2n^4).$$

Hence, in order that $xy+1$ may be a square, one sufficient condition is

$$m^2 = z^2n^4.$$

Therefore $$2z^2 = \frac{2m^2}{n^4} = \frac{(m^2+n^2)+(m^2-n^2)}{[\frac{1}{2}\{(m^2+n^2)-(m^2-n^2)\}]^2}.$$

Again $$x = 2z^2(m^2+n^2) = 2(z^4n^4+z^2n^2),$$

or $$x = 2(p^4+p^2), \text{ if } p = zn.$$

Therefore $$y = 2(p^4-p^2).$$

Example 2. "If thou be expert in mathematics, tell me quickly those two numbers whose sum and difference are squares and whose product is a cube."[1]

That is, solve

$$\left.\begin{aligned} x \pm y &= \text{squares,} \\ xy &= \text{a cube.} \end{aligned}\right\}$$

Bhâskara II says:

"Here let the two numbers be $5z^2$, $4z^2$. *They are assumed such as will make their sum and difference both squares.* Their product is $20z^4$. This must be a cube. Putting it equal to the cube of an optional multiple[2] of $10z$ and removing the common factor z^3 from the sides as before, (we shall ultimately find) the numbers to be 10000, 12500."

[1] *BBi*, p. 56. [2] *G.K*, i. 49.

In general, let us assume, as directed by Bhâskara II,

$$x = (m^2 + n^2)z^2, \quad y = 2mnz^2,$$

which will make $x \pm y$ squares. We have, therefore, only to make

$$2mn(m^2 + n^2)z^4 = \text{a cube.}$$

Let $$2mn(m^2 + n^2)z^4 = p^3z^3.$$

Then $$z = \frac{p^3}{2mn(m^2 + n^2)}.$$

Therefore $$x = \frac{(m^2 + n^2)p^6}{\{2mn(m^2 + n^2)\}^2},$$

$$y = \frac{2mnp^6}{\{2mn(m^2 + n^2)\}^2},$$

where m, n, p are arbitrary.

This general solution has been explicitly stated by Nârâyaṇa thus:

"The square of the cube of an optional number is divided by the square of the product of the two numbers stated above and then severally multiplied by those numbers. (Thus will be obtained) two numbers whose sum and difference are squares and whose product is a cube."[1]

The two numbers stated above[2] are $m^2 + n^2$ and $2mn$ whose sum and difference are squares.

In particular, putting $m = 1$, $n = 2$, $p = 10$, Nârâyaṇa finds $x = 12500$, $y = 10000$. With other values of m, n, p he obtains the values (3165/16, 625/4), (62500/117, 250000/507), (15625/1872, 15625/2028); and observes: "thus by virtue of (the multiplicity of) the optional numbers many values can be found."

[1] *GK*, i. 49. [2] The reference is to rule i. 48.

Example 3. To find numbers such that each of them severally added to a given number becomes a square; and so also the product of every contiguous pair increased by another given number.

For instance, let it be required to find *four* numbers such that

$$x + \alpha = p^2, \qquad xy + \beta = \xi^2,$$
$$y + \alpha = q^2, \qquad yz + \beta = \eta^2,$$
$$z + \alpha = r^2, \qquad zw + \beta = \zeta^2.$$
$$w + \alpha = s^2,$$

The method for the solution of a problem of this kind is indicated in the following rule quoted by Bhâskara II (1150) from an earlier writer, whose name is not known:

"As many multiple (*guṇa*) as the product-interpolator (*vadha-kṣepa*) is of the number-interpolator (*râśi-kṣepa*), with the square-root of that as the common difference are assumed certain numbers; these squared and diminished by the number-interpolator (severally) will be the unknowns."[1]

In applying this method to solve a particular problem, to be stated presently, Bhâskara II observes by way of explanation:

"In these cases, that which being added to an (unknown) number makes it a square is designated as the number-interpolator. The number-interpolator multiplied by the square of the difference of the square-roots pertaining to the numbers, is equal to the product-interpolator. For the product of those two numbers added with the latter certainly becomes a square. The products of two and two contiguous of the square-roots pertaining to the numbers diminished by the

[1] *BBi*, p. 68.

number-interpolator are the square-roots corresponding to the products of the numbers."[1]

Since $x = p^2 - \alpha$, $y = q^2 - \alpha$, we get

$$xy + \beta = (p^2 - \alpha)(q^2 - \alpha) + \beta$$
$$= (pq - \alpha)^2 + \{\beta - \alpha(q - p)^2\}.$$

In order that $xy + \beta$ may be a square, a sufficient condition is

$$\alpha(q - p)^2 = \beta,$$

or $$q = p \pm \sqrt{\beta/\alpha} = p \pm \gamma, \text{ where } \gamma = \sqrt{\beta/\alpha}.$$

Then $xy + \beta = (pq - \alpha)^2.$

Hence $\xi = pq - \alpha.$

Similarly $r = q \pm \gamma, \quad s = r \pm \gamma.$

Thus, it is found that the square-roots p, q, r, s form an A.P. whose common difference is γ $(= \sqrt{\beta/\alpha})$.

Further, we have

$$x = p^2 - \alpha,$$
$$y = (p \pm \gamma)^2 - \alpha,$$
$$z = (p \pm 2\gamma)^2 - \alpha,$$
$$w = (p \pm 3\gamma)^2 - \alpha,$$

as stated in the rule.

These values of the unknowns, it will be easily found, satisfy all the conditions about their products. For

$$xy + \beta = \{p(p \pm \gamma) - \alpha\}^2,$$
$$yz + \beta = \{(p \pm \gamma)(p \pm 2\gamma) - \alpha\}^2,$$
$$zw + \beta = \{(p \pm 2\gamma)(p \pm 3\gamma) - \alpha\}^2.$$

[1] *BBi*, p. 67.

Thus we have

$$\xi = p(p \pm \gamma) - \alpha,$$
$$\eta = (p \pm \gamma)(p \pm 2\gamma) - \alpha,$$
$$\zeta = (p \pm 2\gamma)(p \pm 3\gamma) - \alpha;$$

as stated by Bhâskara II.

It has been observed by him that the above principle is well known in mathematics. But we do not find it in the works anterior to him, which are available to us.

It is noteworthy that the above principle will hold even when all the β's are not equal. For, suppose that in the above instance the second set of conditions is replaced by the following:

$$xy + \beta_1 = \xi^2,$$
$$yz + \beta_2 = \eta^2,$$
$$zw + \beta_3 = \zeta^2.$$

Then, proceeding in the same way, we find that

$$q = p \pm \sqrt{\beta_1/\alpha}, \quad r = q \pm \sqrt{\beta_2/\alpha}, \quad s = r \pm \sqrt{\beta_3/\alpha},$$

and $\qquad \xi = pq - \alpha, \quad \eta = qr - \alpha, \quad \zeta = rs - \alpha.$

It should also be noted that in order that $xy + \beta$ or $p^2q^2 - \alpha(p^2 + q^2) + \alpha^2 + \beta$ may be a square, there may be other values of q besides the one specified above, namely $q = p \pm \sqrt{\beta/\alpha}$. We may, indeed, regard

$$p^2q^2 - \alpha(p^2 + q^2) + \alpha^2 + \beta = \xi^2$$

as an indeterminate equation in q. Since we know one solution of it, namely $q = p \pm \gamma$, $\xi = p(p \pm \gamma) - \alpha$, we can find an infinite number of other solutions by the method of the Square-nature.

Now, suppose that another condition is imposed on the numbers, *viz.*,

$$wx + \beta' = \mu^2.$$

On substituting the values of x and w this condition transforms into

$$p^4 \pm 6\gamma p^3 + (9\gamma^2 + 2\alpha)p^2 \pm 6\alpha\gamma p + \alpha^2 - 9\beta + \beta' = \mu^2,$$

an indeterminate equation of the fourth degree in p.

In the following example and its solution from Bhâskara II we find the application of the above principle:

Example. "What are those four numbers which together with 2 become capable of yielding square-roots; also the products of two and two contiguous of which added by 18 yield square-roots; and which are such that the square-root of the sum of all the roots added by 11 becomes 13. Tell them to me, O algebraist friend."[1]

Solution. "In this example, the product-interpolator is 9 times the number-interpolator. The square-root of 9 is 3. Hence the square-roots corresponding to the numbers will have the common difference 3. Let them be

$$x, x + 3, x + 6, x + 9.$$

"Now the products of two and two contiguous of these minus the number-interpolator are the square-roots pertaining to the products of the numbers as increased by 18. So these square-roots are

$$x^2 + 3x - 2,$$
$$x^2 + 9x + 16,$$
$$x^2 + 15x + 52.$$

"The sum of these and the previous square-roots all together is $3x^2 + 31x + 84$. This added with 11

[1] *BBi*, p. 67.
It will be noticed that by virtue of the last condition the problem becomes, in a way, determinate.

becomes equal to 169. Hence

$$3x^2 + 31x + 95 = 0x^2 + 0x + 169.$$

"Multiplying both sides by 12, superadding 961, and then extracting square-roots, we get

$$6x + 31 = 0x + 43.$$

$$\therefore \quad x = 2.$$

"With the value thus obtained, we get the values of the square-roots pertaining to the numbers to be 2, 5, 8, 11. Subtracting the number-interpolator from the squares of these, we have the (required) numbers as 2, 23, 62, 119."

Example 4. To find two numbers such that

$$\begin{aligned} x - y + k &= u^2, \\ x + y + k &= v^2, \\ x^2 - y^2 + k' &= s^2, \\ x^2 + y^2 + k'' &= t^2. \end{aligned}$$

Bhâskara II says :

"Assume first the value of the square-root pertaining to the difference (of the numbers wanted) to be any unknown with or without an absolute number. The root corresponding to the sum will be equal to the root pertaining to the difference together with the square-root of the quotient of the interpolator of the difference of the squares divided by the interpolator for the sum or difference of the numbers. The squares of these two less their interpolator are the sum and difference of the numbers. From them the two numbers can be found by the rule of concurrence."[1]

[1] *BBi*, pp. 111ff.

That is to say, if w is any rational number, we assume

$$u = w \pm \alpha,$$

where α is an absolute number which may be 0. Then

$$v = (w \pm \alpha) + \sqrt{k'/k}.$$

Now
$$\begin{aligned} x^2 - y^2 + k' &= (x-y)(x+y) + k' \\ &= (u^2 - k)(v^2 - k) + k' \\ &= u^2v^2 - k(u^2 + v^2) + k^2 + k'. \end{aligned}$$

One sufficient condition that the right-hand side may be a square is

$$k(v - u)^2 = k',$$

or

$$v = u + \sqrt{k'/k},$$

which is stated in the rule. Therefore,

$$x - y = (w \pm \alpha)^2 - k,$$
$$x + y = (w \pm \alpha + \sqrt{k'/k})^2 - k.$$

Hence
$$x = \tfrac{1}{2}\{(w \pm \alpha)^2 + (w \pm \alpha + \sqrt{k'/k})^2 - 2k\},$$
$$y = \tfrac{1}{2}\{(w \pm \alpha + \sqrt{k'/k})^2 - (w + \alpha)^2\}.$$

Now, if γ denotes $\sqrt{k'/k}$, we get

$$x^2 + y^2 = u^4 + 2\gamma u^3 + (3\gamma^2 - 2k)u^2 + 2\gamma(\gamma^2 - k)u + \tfrac{1}{2}k^2 + \tfrac{1}{2}(\gamma^2 - k)^2.$$

So it now remains to solve

$$u^4 + 2\gamma u^3 + (3\gamma^2 - 2k)u^2 + 2\gamma(\gamma^2 - k)u + \tfrac{1}{2}k^2 + \tfrac{1}{2}(\gamma^2 - k)^2 + k'' = t^2,$$

which is an indeterminate equation in u.

Applications. We take an illustrative example with its solution from Bhâskara II.

"O thou of fine intelligence, state a pair of numbers, other than 7 and 6, whose sum and difference (severally) added with 3 are squares; the sum of their squares decreased by 4 and the difference of the squares increased by 12 are also squares; half their product together with the smaller one is a cube; again the sum of all the roots plus 2 is a square."[1]

That is to say, if $x > y$, we have to solve

$$x - y + 3 = u^2,$$
$$x + y + 3 = v^2,$$
$$x^2 - y^2 + 12 = s^2,$$
$$x^2 + y^2 - 4 = t^2,$$
$$\tfrac{1}{2}xy + y = p^3,$$
$$u + v + s + t + p + 2 = q^2.$$

This problem has been solved in two ways:

First Method. As directed in the above rule, assume

$$u = w - 1.$$

Then $\quad x - y = (w-1)^2 - 3 = w^2 - 2w - 2,$

$$x + y = (\overline{w-1} + 2)^2 - 3 = w^2 + 2w - 2.$$

Therefore $\quad x = w^2 - 2, \quad y = 2w.$

Now, we find that

$$x^2 - y^2 + 12 = (w^2 - 4)^2,$$
$$x^2 + y^2 - 4 = w^4,$$
$$\tfrac{1}{2}xy + y = w^3.$$

So all the equations except the last one are already satisfied. This remaining equation now reduces to

$$2w^2 + 3w - 2 = q^2.$$

Completing the square on the left-hand side of this equation, we get

$$(4w + 3)^2 = 8q^2 + 25.$$

[1] *BBi*, p. 115.

By the method of the Square-nature its solutions are

$$\left.\begin{array}{r} q = 5 \\ 4w + 3 = 15 \end{array}\right\}, \quad \left.\begin{array}{r} q = 175 \\ 4w + 3 = 495 \end{array}\right\}, \ldots$$

Therefore $\quad w = 3, 123, \ldots$

Hence the values of (x, y) are $(7, 6), (15127, 246), \ldots$

Second Method. Or assume[1]

$$x - y + 3 = w^2,$$

then $\quad x + y + 3 = w^2 + 4w + 4 = (w + 2)^2.$

Whence $\quad x = w^2 + 2w - 1, \quad y = 2w + 2.$

Now, we find that

$$x^2 - y^2 + 12 = (w^2 + 2w - 3)^2,$$
$$x^2 + y^2 - 4 = (w^2 + 2w + 1)^2,$$
$$\tfrac{1}{2}xy + y = (w + 1)^3.$$

Then the remaining condition reduces to

$$2w^2 + 7w + 3 = q^2.$$

Completing the square on the left-hand side, we get

$$(4w + 7)^2 = 8q^2 + 25.$$

Whence by the method of the Square-nature, we get

$$\left.\begin{array}{r} q = 5 \\ 4w + 7 = 15 \end{array}\right\}, \quad \left.\begin{array}{r} q = 175 \\ 4w + 7 = 495 \end{array}\right\}, \ldots$$

Therefore $\quad w = 2, 122, \ldots$

Hence $\quad (x, y) = (7, 6), (15127, 246), \ldots$

Another very interesting example which has been borrowed by Bhâskara II from an earlier writer is the following:[2]

[1] This is clearly equivalent to the supposition, $u = w$, $v = w + 2$.

[2] The text is *kasyâpyudâharaṇaṃ* ("the example of some one"). This observation appears to indicate that this particular example was borrowed by Bhâskara II from a secondary source; its primary source was not known to him.

"Tell me quickly, O sound algebraist, two numbers, excepting 6 and 8, which are such that the cube-root of half the sum of their product and the smaller one, the square-root of the sum of their squares, the square-roots of the sum and difference of them (each) increased by 2, and of the difference of their squares plus 8, all being added together, will be capable of yielding a square-root."[1]

That is to say, if $x > y$, we have to solve

$$\sqrt[3]{\tfrac{1}{2}(xy + y)} + \sqrt{x^2 + y^2} + \sqrt{x^2 - y^2 + 8}$$
$$+ \sqrt{x + y + 2} + \sqrt{x - y + 2} = q^2.$$

In every instance of this kind, remarks Bhâskara II, "the values of the two unknown numbers should be so assumed in terms of another unknown that all the stipulated conditions will be satisfied." In other words, the equation will have to be resolved into a number of other equations all of which have to be satisfied simultaneously. Thus we shall have to solve

$$x - y + 2 = u^2,$$
$$x + y + 2 = v^2,$$
$$x^2 - y^2 + 8 = s^2,$$
$$x^2 + y^2 = t^2,$$
$$\tfrac{1}{2}(xy + y) = p^3,$$
$$u + v + s + t + p = q^2.$$

The last equation represents the original one.

There have been indicated several methods of solving these equations.

(*i*) Set $x = w^2 - 1$, $y = 2w$; then we find that

$$x - y + 2 = (w - 1)^2,$$
$$x + y + 2 = (w + 1)^2,$$

[1] *BBi*, p. 110.

$$x^2 - y^2 + 8 = (w^2 - 3)^2,$$
$$x^2 + y^2 = (w^2 + 1)^2,$$
$$\tfrac{1}{2}(xy + y) = w^3.$$

So all the equations except the last one are identically satisfied. This last equation now becomes

$$2w^2 + 3w - 2 = q^2.$$

Completing the square on the left-hand side, we get

$$(4w + 3)^2 = 8q^2 + 25.$$

Solutions of this are

$$\left.\begin{matrix} q = 5 \\ 4w + 3 = 15 \end{matrix}\right\}, \quad \left.\begin{matrix} q = 30 \\ 4w + 3 = 85 \end{matrix}\right\}, \quad \left.\begin{matrix} q = 175 \\ 4w + 3 = 495 \end{matrix}\right\}, \ldots$$

Therefore, we have the solutions of our problem as

$$(x, y) = (8, 6), (1677/4, 41), (15128, 246), \ldots$$

Or set

$$(ii) \quad \begin{cases} x = w^2 + 2w, \\ y = 2w + 2; \end{cases}$$

$$(iii) \quad \begin{cases} x = w^2 - 2w, \\ y = 2w - 2; \end{cases}$$

$$\text{or} \quad (iv) \quad \begin{cases} x = w^2 + 4w + 3, \\ y = 2w + 4. \end{cases}$$

In conclusion Bhâskara II remarks: "Thus there may be a thousandfold artifices; since they are hidden to the dull, a few of them have been indicated here out of compassion for them."[1]

It will be noticed that in devising the various artifices noted above for the solution of the problem, Bhâskara II has been in each case guided by the result that if $u = w \pm \alpha$, then, $v = w \pm \alpha + \sqrt{k'/k}$. He has simply taken different values of α in the different cases.

[1] *BBi*, p. 110.

28. SOLUTION OF $axy = bx + cy + d$

Bakhshâlî Treatise. The earliest instance of a quadratic indeterminate equation of the type $axy = bx + cy + d$, in Hindu mathematics occurs in the Bakhshâlî Treatise (*c.* 200).[1] The text is very mutilated. But the example that is preserved is

$$xy = 3x + 4y \mp 1,$$

of which the solutions preserved are

$$x = \frac{3 \cdot 4 - 1}{1} + 4 = 15,$$

$$y = 1 + 3 = 4;$$

and

$$x = 1 + 4 = 5,$$

$$y = \frac{3 \cdot 4 + 1}{1} + 3 = 16.$$

Hence, in general, the solutions of the equation

$$xy = bx + cy + d,$$

which appear to have been given are:

$$\left.\begin{aligned} x &= \frac{bc + d}{m} + c, \\ y &= m + b; \end{aligned}\right\} \quad \text{or} \quad \left.\begin{aligned} x &= m + c, \\ y &= \frac{bc + d}{m} + b; \end{aligned}\right\}$$

where m is an arbitrary number.

An Unknown Author's Rule. Brahmagupta (628) has described the following method taken from an author who is not known now.[2]

[1] *BMs*, Folio 27, recto; compare also Kaye's Introduction §82.

[2] Pṛthûdakasvâmî (860) says that the method is due to a writer other than Brahmagupta. This is further corroborated by Brahmagupta's strictures on it (*vide infra*, p. 299).

"The product of the coefficient of the factum and the absolute number together with the product of the coefficients of the unknowns is divided by an optional number. Of the optional number and the quotient obtained, the greater is added to the lesser (of the coefficients of the unknowns) and the lesser to the greater (of the coefficients), and (the sums) are divided by the coefficient of the factum. (The results will be values of the unknowns) in the reverse order."[1]

As has been observed by Pṛthûdakasvâmî, this rule is to be applied to an equation containing the factum after it has been prepared by transposing the factum term to one side and the absolute term together with the simple unknown terms to the other. Then the solutions will be, m being an arbitrary rational number,

$$x = \frac{1}{a}(m + c),$$

$$y = \frac{1}{a}\left(\frac{ad + bc}{m} + b\right),$$

if $b > c$ and $m > \frac{ad + bc}{m}$. If these conditions be reversed then x and y will have their values interchanged.

The *rationale* of the above solutions can be easily shown to be as follows:

$$axy = bx + cy + d,$$

or $$a^2xy - abx - acy = ad,$$

or $$(ax - c)(ay - b) = ad + bc.$$

Suppose $ax - c = m$, a rational number;

then $$ay - b = \frac{ad + bc}{m}.$$

[1] *BrSpSi*, xviii, 60.

Therefore $$x = \frac{1}{a}(m + c),$$

$$y = \frac{1}{a}\left(\frac{ad + bc}{m} + b\right).$$

Or, we may put $ay - b = m$;

then we shall have $$ax - c = \frac{ad + bc}{m};$$

whence $$x = \frac{1}{a}\left(\frac{ad + bc}{m} + c\right),$$

$$y = \frac{1}{a}(m + b).$$

It will thus be found that the restrictive condition of adding the greater and lesser of the numbers m and $(ad + bc)/m$ to the lesser and greater of the numbers b and c respectively as adumbrated in the above rule is quite unnecessary.

Brahmagupta's Rule. Brahmagupta gives the following rule for the solution of a quadratic indeterminate equation involving a factum:

"With the exception of an optional unknown, assume arbitrary values for the rest of the unknowns, the product of which forms the factum. The sum of the products of these (assumed values) and the (respective) coefficients of the unknowns will be absolute quantities. The continued products of the assumed values and of the coefficient of the factum will be the coefficient of the optionally (left out) unknown. Thus the solution is effected without forming an equation of the factum. Why then was it done so?"[1]

The reference in the latter portion of this rule is to the method of the unknown writer. The principle

[1] *BrSpSi*, xviii. 62-3, *vide supra*, p. 297.

underlying Brahmagupta's method is to reduce, like the Greek Diophantus (*c.* 275),[1] the given indeterminate equation to a simple determinate one by assuming arbitrary values for all the unknowns except one. So it is undoubtedly inferior to the earlier method. Brahmagupta gives the following illustrative example :

"On subtracting from the product of signs and degrees of the sun, three and four times (respectively) those quantities, ninety is obtained. Determining the sun within a year (one can pass as a proficient) mathematician."[2]

If x denotes the signs and y the degrees of the sun, then the equation is

$$xy - 3x - 4y = 90.$$

Thus this problem, as that of Bhâskara II (*infra*), appears to have some relation with that of the Bakhshâlî work. Pṛthûdakasvâmî solves it in two ways. *Firstly*, he assumes the arbitrary number to be 17, then

$$x = \frac{1}{1}\left(\frac{90.1 + 3.4}{17} + 4\right) = 10,$$

$$y = \frac{1}{1}(17 + 3) = 20.$$

Secondly, he assumes arbitrarily $y = 20$. On substituting this value in the above equation, it reduces to

$$20x - 3x = 170\,;$$

whence $x = 10$.

Mahâvîra's Rule. Mahâvîra (850) has not treated equations of this type. There are, however, two problems in his *Gaṇita-sâra-saṁgraha* which involve similar equations. One of them is to find the increase or

[1] Heath, *Diophantus*, pp. 192-4, 262.

[2] *BrSpSi*, xviii. 61.

decrease of two numbers (a, b) so that the product of the resulting numbers will be equal to another optionally given number (d). Thus we are to solve

$$(a \pm x)(b \pm y) = d,$$

or

$$xy \pm (bx + ay) = d - ab.$$

The rule given for solving this is:

"The difference between the product of the given numbers and the optional number is put down at two places. It is divided (at one place) by one of the given numbers increased by unity and (at the other) by the optional number increased by the other given number. These will give in the reverse order the values of the quantities to be added or subtracted."[1]

That is to say,

$$\left.\begin{aligned} x &= \frac{d \sim ab}{d+b} \\ y &= \frac{d \sim ab}{a+1} \end{aligned}\right\}, \quad \text{or} \quad \left.\begin{aligned} x &= \frac{d \sim ab}{b+1} \\ y &= \frac{d \sim ab}{d+a} \end{aligned}\right\}$$

Thus the solutions given by Mahâvîra are much cramped. The other problem considered by him is to separate the capital, interest and time when their sum is given: If x be the capital invested and y the period of time in months, then the interest will be mxy, where m is the rate of interest per month. Then the problem is to solve

$$mxy + x + y = p.$$

Mahâvîra solves this equation by assuming arbitrary values for y.[2]

[1] *GSS*, vi. 284. [2] *GSS*, vi. 35.

Śrîpati's Rule. Śrîpati (1039) gives the following rule:

"Remove the factums from one side, the (simple) unknowns and the absolute numbers from the other. The product of the coefficients of the unknowns being added to the product of the absolute quantity and the coefficient of the factum, (the sum) is divided by an optional number. The quotient and the divisor should be added arbitrarily to the greater or smaller of the coefficients of the unknowns. These divided by the coefficient of the factum will be the values of the unknowns in the reverse order."[1]

$$i.e.,\ \left.\begin{aligned} x &= \frac{1}{a}(m+c) \\ y &= \frac{1}{a}\left(\frac{ad+bc}{m}+b\right) \end{aligned}\right\},\qquad \left.\begin{aligned} x &= \frac{1}{a}\left(\frac{ad+bc}{m}+c\right) \\ y &= \frac{1}{a}(m+b) \end{aligned}\right\},$$

where m is arbitrary.

Bhâskara II's Rule. Bhâskara II (1150) has given two rules for the solution of a quadratic indeterminate equation containing the product of the unknowns. His first method is the same as that of Brahmagupta:

"Leaving one unknown quantity optionally chosen, the values of the other should be assumed arbitrarily according to convenience. The factum will thus be reduced and the required solution can then be obtained by the first method of analysis."[2]

Bhâskara's aim was to obtain *integral* solutions. The above method is, however, not convenient for the purpose. He observes:

"On assuming in this way an arbitrary known value for one of the unknowns, the *integral* values of the

[1] *SiŚe*, xiv. 20-1. [2] *BBi*, p. 123.

two unknowns can be obtained with much difficulty."[1] So he describes a second method "by which they can be obtained with little difficulty."

"Transposing the factum from one side chosen at pleasure, and the (simple) unknowns and the absolute number from the other side (of the equation), and then dividing both the sides by the coefficient of the factum, the product of the coefficients of the unknowns together with the absolute number is divided by an optional number. The optional number and that quotient should be increased or diminished by the coefficients of the unknowns at pleasure. They (results thus obtained) should be known as the values of the two unknowns reciprocally."[2]

This rule has been elucidated by the author thus:

"From one of the two equal sides the factum being removed, and from the other the unknowns and the absolute number; then dividing the two sides by the coefficients of the factum, the product of the coefficients of the unknowns on the other side added to the absolute number, is divided by an optional number. The optional number and the quotient being arbitrarily added to the coefficients of the unknowns, should be known as the values of the unknowns in the reciprocal order. That is, the one to which the coefficient of the *kâlaka* (the second unknown) is added, will be the value of the *yâvat-tâvat* (the first unknown) and the one to which the coefficient of the *yâvat-tâvat* is added, will be the value of the *kâlaka*. But if, after that has been done, owing to the magnitude, the statements (of the problem) are not fulfilled, then

1 "Evamekasmin vyakte râśau kalpite sati bahûnâyâsenabhinnau râśî jñâyete"—*BBi*, p. 124.

2 *BBi*, pp. 124f.

from the optional number and the quotient, the coefficients of the unknowns should be subtracted, and (the remainders) will be the values of the unknowns in the reciprocal order."

Thus Bhâskara's solutions are

$$\left.\begin{aligned} x &= \frac{c}{a} \pm m' \\ y &= \frac{b}{a} \pm n' \end{aligned}\right\} \qquad \left.\begin{aligned} x &= \frac{c}{a} \pm n' \\ y &= \frac{b}{a} \pm m' \end{aligned}\right\}$$

where m' is any arbitrary number and $n' = \frac{1}{m'}\left(\frac{bc}{a^2} + \frac{d}{a}\right)$.

The *rationale* of these solutions is as follows :

$$axy = bx + cy + d,$$

or $$xy - \frac{b}{a}x - \frac{c}{a}y = \frac{d}{a},$$

or $$\left(x - \frac{c}{a}\right)\left(y - \frac{b}{a}\right) = \frac{d}{a} + \frac{bc}{a^2} = m'n', \text{ say.}$$

Then, either

$$\left.\begin{aligned} x - \frac{c}{a} &= \pm m' \\ y - \frac{b}{a} &= \pm n' \end{aligned}\right\} \quad \text{or} \quad \left.\begin{aligned} x - \frac{c}{a} &= \pm n' \\ y - \frac{b}{a} &= \pm m' \end{aligned}\right\};$$

whence the solutions.

Bhâskara's Proofs. The same *rationale* of the above solutions has been given also by Bhâskara II with the help of the following illustrative example. He observes that the proof "is twofold in every case : one geometrical (*kṣetragata*), the other algebraic (*râśigata*)."[1]

Example. "The sum of two numbers multiplied by four and three, added by two is equal to the product

[1] *BBi*, p. 125.

of those numbers. Tell me, if thou knowest, those two numbers."[1]

Solution. "Having performed the operations as stated, the sides are

$$xy = 4x + 3y + 2.$$

The product of the coefficients of the unknowns plus the absolute term is 14. Dividing this by an optional number (say) unity, the optional number and the quotient are 1, 14. To these being arbitrarily added 4, 3, the coefficients of the unknowns, the values of (x, y) are (4, 18) or (17, 5). (Dividing) by (the optional number) 2, (other values will be) (5, 11) or (10, 6.)"[2]

Geometrical Proof. "The second side of the equation is equal to the factum. But the factum is the area of an oblong quadrilateral of which the base and upright are the unknown quantities. Within this figure (Fig. 15) exist four x's, three y's and the absolute number 2. From this figure on taking off four x's and y minus four multiplied by its own coefficient, (*i.e.*, 3), it becomes this (Fig. 16).

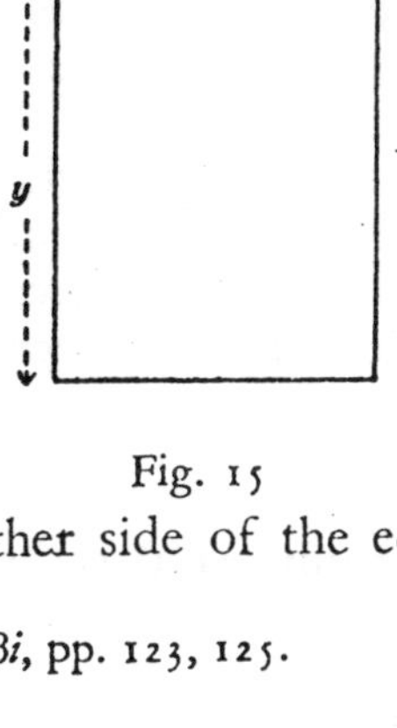

Fig. 15

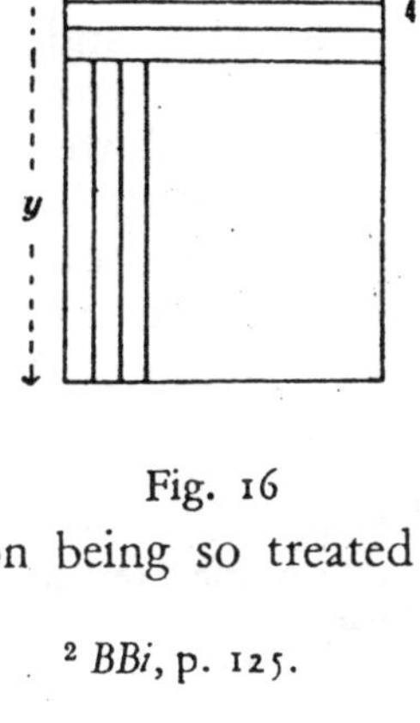

Fig. 16

The other side of the equation being so treated there

[1] *BBi*, pp. 123, 125.

[2] *BBi*, p. 125.

results 14. This must be the area of the figure remaining at the corner (see Fig. 16) within the rectangle representing the factum, and is the product of its base and upright. But these are (still) to be known here. Therefore, assuming an optional number for the base, the upright will be obtained on dividing the area 14 by it. One of these, base and upright, being increased by 4, the coefficient of x will be the upright of the figure representing the factum, because when four x's were separated from the factum-figure, its upright was lessened by 4. Similarly the other being increased by 3, the coefficient of y, will be the base They are precisely the values of x and y."[1]

Algebraic Proof "This is also geometrical in origin. In this the values of the base and upright of the smaller rectangle within the rectangle whose base and upright are x and y respectively, are assumed to be two other unknowns u and v.[2] One of them being increased by the coefficient of x will be the value of the upright of the outer figure and the other being increased by the coefficient of y will be taken to be the value of the base of the outer figure. Thus $y = u + 4$, $x = v + 3$. Substituting these values of the unknowns x, y, on both sides of the equation, the upper side will be $3u + 4v + 26$ and the factum side will be $uv + 3u + 4v + 12$. On making perfect clearance between these sides, the lower side becomes uv and the upper side 14. This is the area of that inner rectangle and it is equal to the product of the coefficients of the unknowns plus the absolute number. How the values of the unknowns are to be thence deduced, has been already explained."[3]

[1] *BBi*, p. 126.

[2] In the original text they are respectively *nî* (for *nîlaka*) and *pî* (for *pîtaka*).

[3] *BBi*, p. 127.

Bhâskara II further observes :

"Thus the proof of the solution of the factum has been shown to be of two kinds. What has been said before—the product of the coefficients of the unknowns together with the absolute number is equal to the area of another rectangle inside the rectangle representing the factum and lying at a corner—is sometimes otherwise. For, when the coefficients of the unknowns are negative, the factum-rectangle will be inside the other rectangle at one corner ; and when the coefficients of the unknowns are greater than the base and upright of the factum-rectangle, and are positive, the other will be outside the factum rectangle and at a corner, as (Figs. 17, 18).

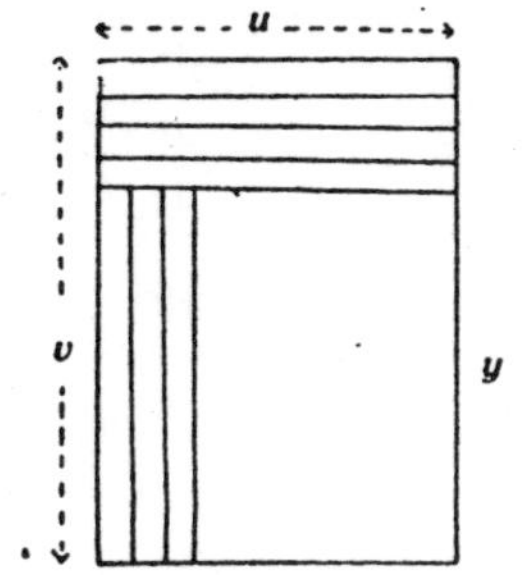

Fig. 17

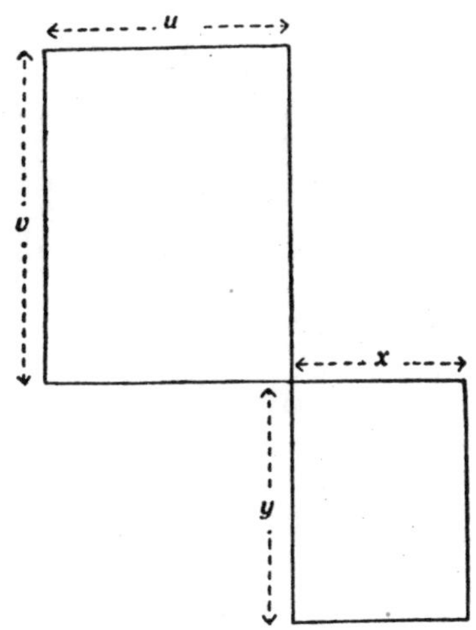

Fig. 18

When it is so, the coefficients of the unknowns lessened by the optional number and the quotient, will be the values of x and y."[1]

[1] *BBi*, p. 127.

INDEX